ANNALS *of* THE NEW YORK ACADEMY OF SCIENCES

EDITOR-IN-CHIEF
Douglas Braaten

ASSOCIATE EDITOR
Rebecca E. Cooney

PROJECT MANAGER
Steven E. Bohall

EDITORIAL ADMINISTRATOR
Daniel J. Becker

Artwork and design by Ash Ayman Shairzay

The New York Academy of Sciences
7 World Trade Center
250 Greenwich Street, 40th Floor
New York, NY 10007-2157

annals@nyas.org
www.nyas.org/annals

About the Blavatnik Family Foundation

The Blavatnik Family Foundation is an active supporter of many leading educational, scientific, cultural, and charitable institutions in the United States, the United Kingdom, Israel, and throughout the world. Recipients of foundation support include Oxford University, Harvard University, Tel Aviv University, the Royal Opera House, the Hermitage, the National Portrait Gallery, the British Museum, the National Gallery of Art, the Metropolitan Museum of Art, the New York Academy of Sciences, the White Nights Foundation of America, numerous Jewish charitable organizations, and countless other philanthropic institutions. The foundation is headed by Len Blavatnik, an American industrialist, who is the founder and chairman of Access Industries, a privately held industrial group with global interests in natural resources and chemicals, media and telecommunications, and real estate.

BLAVATNIK FAMILY FOUNDATION

The New York
Academy of Sciences

Published by Blackwell Publishing
On behalf of the New York Academy of Sciences

Boston, Massachusetts
2012

ANNALS *of* THE NEW YORK ACADEMY OF SCIENCES

VOLUME
1260

ISSUE

Blavatnik Awards for Young Scientists 2011

The New York Academy of Sciences Blavatnik Awards for Young Scientists acknowledge and celebrate the excellence of our most noteworthy young scientists and engineers in New York, New Jersey, and Connecticut. The Awards recognize highly innovative, impactful, and interdisciplinary accomplishments in the life sciences, physical sciences, mathematics, and engineering with unrestricted financial prizes for both finalists and awardees.

TABLE OF CONTENTS

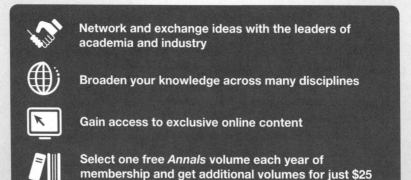

Ann. N.Y. Acad. Sci. ISSN 0077-8923

ANNALS OF THE NEW YORK ACADEMY OF SCIENCES

Issue: *Blavatnik Awards for Young Scientists*

A crosstalk between bone and gonads

Franck Oury

Department of Genetics and Development, Columbia University, New York, New York

Address for correspondence: Franck Oury, Ph.D., Department of Genetics and Development, 701 West 168th Street, HHSC1602, New York, NY 10032. fo2133@columbia.edu

The sex steroid hormones testosterone and estrogen are essential determinants not only of reproductive functions but also for bone growth and the maintenance of skeletal integrity. The importance of this latter form of regulation is best exemplified by the fact that gonadal failure triggers bone loss in both genders and causes osteoporosis in postmenauposal women. Traditionally, bone physiology is studied with the view that the skeleton is simply a recipient of hormonal inputs. However, a richer picture of bone physiology has recently emerged, and it is now clear that the skeleton is an endocrine organ itself. This is particularly relevant to the interplay between bone and gonads because genetics and biochemical evidence have established that bone, via the osteoblast-derived hormone osteocalcin, promotes testosterone biosynthesis. This review will present the mechanism of action of osteocalcin and will discuss the implications of this novel regulation.

Keywords: sex steroids; gonads; bone; osteocalcin; Gprc6a

Introduction

The sex steroid hormones estrogen and androgen are multifunctional physiological determinants. Acting through nuclear hormonal receptors, testosterone promotes the development and maintenance of male reproductive functions, while estrogen favors ovarian function, ovocyte maturation, and ovulation. In addition to their reproductive functions, sex steroid hormones are also powerful regulators of bone physiology. Testosterone and estrogen influence positively the growth, maturation, and maintenance of the female and male skeleton. The biological importance of this regulation is best exemplified by the fact that gonadal failure triggers bone loss in both genders and causes osteoporosis in postmenopausal women.[1-6]

Bone has emerged in recent years as an endocrine organ of growing importance. This was first established by showing that energy metabolism is influenced by a bone-derived hormone called osteocalcin.[7] More recently, it was shown that osteocalcin also favors testosterone biosynthesis.[8] It is this latter aspect of osteocalcin biology that will be reviewed here.

Sex steroid hormone effects on bone mass

Because of the essential role of the skeleton in locomotion, support, and protection of the body, it is critical for vertebrates to maintain high bone quality. This is achieved through the unique ability of bone to constantly renew itself via the succession of two functions: bone resorption by osteoclasts, followed by bone formation by osteoblasts.[9-12] The concerted action of these two cell types defines bone modeling during childhood and remodeling during adulthood.

The regulation of bone (re)modeling is complex and involves mechanical stimuli, locally produced factors, and hormones. Among these hormones, sex steroids play a crucial role during the bone growth spurts of puberty, and for the maintenance of bone mass.

Sex steroids regulate bone growth

Skeletal size and volume are similar in prepubertal girls and boys.[13,14] However, gender differences in bone growth become apparent during puberty, with men reaching higher peak bone mass. The male skeleton is characterized by larger bone size and stronger skeleton. These skeletal gender

doi: 10.1111/j.1749-6632.2011.06360.x

differences are attributed to a stimulatory androgen action on periosteal bone formation in men versus an inhibitory estrogen-related action in women.[13,15–18] This androgen effect is best illustrated by the fact that prepubertal hypogonadism is associated with low bone mineral density at puberty, while administration of testosterone before epiphyseal closure leads to increases in bone mass.[19–21] Moreover, there is evidence that androgens also have effects on the peak of bone mass in women since an excess of androgen in women is associated with higher bone mineral density.[22–24] However, it was recently shown that androgens are not the only factors involved in this process. Other hormones negatively regulated by estrogen, such as growth hormone (GH) and insulin-like growth factor 1 (IGF1), may further contribute to the development of the skeletal sexual dimorphism.[25–27]

Sex steroids regulate bone mass maintenance

Testosterone and estrogens are also implicated in the maintenance of bone mass integrity during adulthood in the female and male skeleton; they are crucial determinants for maintaining bone homeostasis. Indeed, the absence or decrease of either testosterone or estrogen levels, with age or in gonadal dysfunction, leads to a decrease in bone mass and increases markedly the risk of osteoporosis. As a matter of fact, the loss of ovarian function at menopause is the most important factor for the development of osteoporosis.[1–5] Likewise, loss of androgens in males following castration or a decrease in androgen levels related to aging, have the same dramatic effect on the skeleton.[4,6,28,29] However, the dichotomy associating androgens and estrogens as a pure "male" and "female" hormones, respectively, has recently been reconsidered.[27] Recent evidence showed that estrogens may also have crucial roles in the maintenance of bone mass accrual and skeletal homeostasis in elderly men.[30–34] This novel notion suggests that the role of sex steroids in the maintenance of bone mass is more complex than previously anticipated.

Estrogen and androgen mode of action in skeleton

Estrogens actively suppress bone turnover and maintain balanced rates of both bone formation and bone resorption (Fig. 1). At the cellular level, estrogens affect the generation, lifespan, and functional activity of both osteoclasts and osteoblasts. They decrease osteoclast formation and activity, while they increase osteoclast apoptosis.[35,36] The action of estrogens on osteoblasts is less clear and its investigation has produced conflicting results. While the majority of the studies indicate that estrogen suppresses osteoblast differentiation, in contrast, some evidence suggest that estrogens may also increase osteoblast differentiation, proliferation, and function.[6,37,38] These opposing findings are most probably due to different potent actions of estrogen depending on the stage of osteoblast differentiation.

At the molecular level, estrogens decrease the production of cytokines inhibiting osteoclast apoptosis such as IL-1, IL-6, TNF-α, and M-CSF, and downregulate the expression of NF-κB–activated gene, a suppressor of apoptosis.[4,36,39–42] In contrast, estrogens favor expression of TGF-α, a direct inhibitor of osteoclast activity and an activator of osteoclast apoptosis.[43–45] Estrogens can also directly suppress bone resorption. They enhance production of OPG and decrease RANKL expression by osteoblasts, an inhibitor and an activator of osteoclast activity, respectively.[46] Lastly, estrogens also suppress osteoclast formation by inducing osteoclast apoptosis through an activation of Fas/FasL signaling.[3] They favor expression of Fas ligand (*FasL*), a gene that belongs to the tumor necrosis factor (TNF) family, in osteoclasts and osteoblasts.[3,47] Therefore, the activity of estrogens on bone is an important physiological mechanism to maintain bone mass during adulthood (Fig. 1).

Androgens have potent effects on osteoblast formation, and those are differential whether they act on trabecular or periosteal bone. While androgens maintain trabecular bone mass and integrity, they favor periosteal bone formation in men.[6] At the cellular level, testosterone increases the lifespan of osteoblasts by decreasing apoptosis, mainly through its action on IL-6 production (Fig. 1).[42,48] Furthermore, androgens stimulate the proliferation of osteoblast progenitors and the differentiation of mature osteoblasts.[42,49–51] The molecular mechanisms of androgen action on bone cells are less well described than for estrogens. However, some evidence indicates that they favor osteoblast proliferation and differentiation by increasing TGF mRNA and responsiveness to FGF and IGF-II.[4,44,45] Androgen

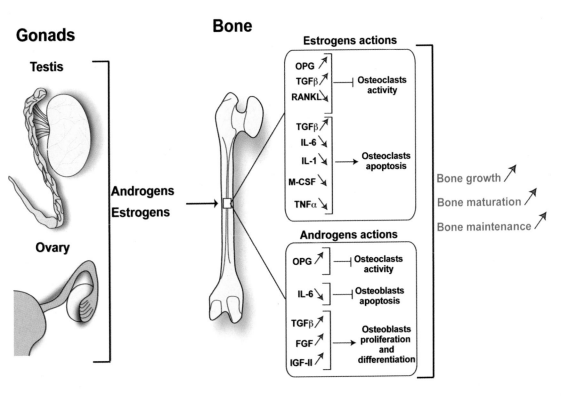

Figure 1. Sex steroid hormones regulate bone growth, maturation, and maintenance. Sex steroid hormones, testosterone and estrogen, secreted by gonads (testis or ovary) play a crucial role during skeletal growth, maturation, and maintenance in both males and females. This figure summarizes the major targets of estrogens and androgens in bone cells. Their activation or repression is associated with effects on apoptosis and/or proliferation of osteoblasts and/or osteoclasts.

may also decrease osteoclast formation and bone resorption through increased production of OPG by osteoblasts.[46] The net result of these functions of testosterone is to favor bone formation (Fig. 1). Moreover, testosterone may influence different stages of osteoblast differentiation and may act on osteoblasts differently than estrogen at various skeletal localizations. This last aspect of testosterone on bone formation is also important for the skeletal sexual dimorphism.[13]

Mediators of sex steroid hormone actions on bone cells

Classical receptors for estrogens (ERs) or androgens (AR) are expressed in bone cells, indicating that the sex steroid hormones may influence skeletal physiology, at least in part, by acting directly on bone cells.[27,52,53] The analysis of ERα (αER-KO), ERβ (βER-KO), double ER (DER-KO), and AR (Ar-KO) mutated mice showed some skeletal abnormalities.[54–56] Yet ER-deficient mouse models have been confounded by highly elevated levels of serum sex steroids and a weak phenotype in comparison to the one observed in estrogen-deficient mouse models.[4,57] In addition, it has been shown that sex steroid hormones may also act indirectly in the context of nongenomic sex steroid signaling. Androgens and estrogens can transmit antiapoptotic effects on osteoblasts *in vitro* with a similar efficiency via either AR or ERs, irrespective of whether the ligand is an androgen or an estrogen.[42,49,50] In conclusion, taken together, these data suggest that the mechanisms mediating estrogen and androgen function in the bone via ERs and AR are not fully elucidated.

The skeleton returns the favor to the gonads

Recently, multiple studies have shown that bone is an endocrine organ favoring whole-body glucose homeostasis and energy expenditure, both in mice and in humans.[7,52,58,59] These novel functions of bone are mediated by an osteoblast-specific secreted hormone osteocalcin that, when under carboxylated, favors cell proliferation, insulin secretion, and insulin sensitivity in muscle, liver, and white adipose tissue.

The hormonal functions of osteocalcin have raised multiple questions of great biological and medical importance. Chief among them is to elucidate the signaling events triggered by this hormone in target cells. A second question, with even broader implications, is to determine whether osteocalcin, like many other hormones, has functions in addition to those exerted on energy metabolism. The well-known regulation of bone remodeling by gonads provides an ideal setting to address the aforementioned question.

As mentioned previously, that menopause favors bone loss is well established. What this medical observation means biologically is that gonads, mostly through sex steroid hormones, affect the functions of bone cells, an aspect of bone physiology that has been discussed above. According to the general principle of feedback control, what the regulation of bone mass accrual by gonads also suggests is that bone may affect the reproductive functions in one or both genders. Verifying this hypothesis was of great conceptual importance, as it would further enhance the emerging notion of the importance of bone as an endocrine organ.

The first hint that this hypothesis could be true comes from *ex vivo* cell assays. Indeed, these studies showed that a factor secreted by osteoblasts, but not by other cells of mesodermal origin, could markedly increase testosterone production by testis explants and primary Leydig cells.[8] This novel role of osteoblasts was verified recently *in vivo*.[60] Interestingly, this effect was limited to the male; osteoblasts did not stimulate testosterone or estrogen production in the ovaries.

Osteocalcin, a new player in the regulation of testosterone production

Since the hormone secreted by osteoblasts, osteocalcin, has an important role in the regulation of energy metabolism and glucose homeostasis, it was hypothesized that it may also affect testosterone production. Testing this hypothesis *in vivo* was greatly helped by the fact that while female osteocalcin-deficient mice ($Ocn^{-/-}$) were normally fertile, the male mutant mice were rather poor breeders. The demonstration of a reproductive func-

tion of osteocalcin in male mice was greatly helped by the availability of both gain ($Esp^{-/-}$) and loss-of-function ($Ocn^{-/-}$) mutations for osteocalcin functions.[7] Osteocalcin-deficient mice showed a decrease in testes, epidydimedes, and seminal vesicles weights, the opposite was seen in male $Esp^{-/-}$ mice. $Ocn^{-/-}$ male mice showed a 50% decrease in sperm count, whereas $Esp^{-/-}$ male mice showed a 30% increase in this parameter. Moreover, Leydig cell maturation appeared to be halted in the absence of osteocalcin. These features suggested that osteocalcin could favor testosterone synthesis. This hypothesis was verified by coculture assays, since the supernatants of WT but not of $Ocn^{-/-}$ osteoblasts in culture increased testosterone production by Leydig cells of the testes. Accordingly, circulating testosterone levels are low in $Ocn^{-/-}$ and high in $Esp^{-/-}$ male mice.[8] Taken together, these experiments established that osteocalcin is a bone-derived hormone favoring fertility in male mice by promoting testosterone production by Leydig cells (Fig. 2). In other words, it verified that, for at least one gender, there is an endocrine regulation of reproduction by the skeleton. It also suggested that there might be differences between males and females in the regulation of this function.

Osteocalcin mode of action in Leydig cells

The identification of a novel hormone immediately raises the question of its mechanism of action. A prerequisite to answering this question is to identify the receptor to which this hormone would bind specifically on its target cells.

In the first step to address this question, we searched for the signal transduction pathway affected by osteocalcin in two target cells, the β-cell of the pancreas and the Leydig cell of the testis. This approach identified the production of cAMP as the only intracellular signaling event reproducibly triggered by osteocalcin in these two cell types, suggesting that the osteocalcin receptor would be a G protein–coupled receptor (GPCR) linked to adenylate cyclase.

Out of more than 100 orphan GPCRs analyzed, 22 of them were more expressed in testes than in ovaries and only four were expressed predominantly, or only, in Leydig cells.[8,61] One of these four orphan GPCRs, Gprc6a, was a particularly good candidate to be an osteocalcin receptor since its inactivation in mice results in metabolic and reproduction

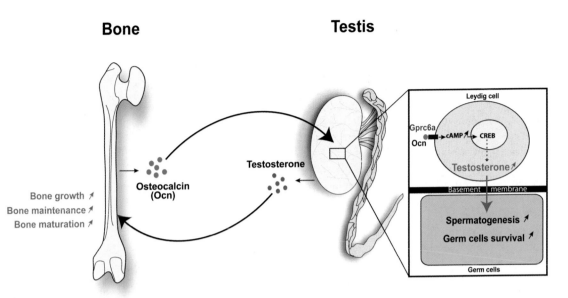

Figure 2. Bone endocrine regulation of testosterone biosynthesis by Leydig cells in testis. Bone via osteocalcin, an osteoblast-derived hormone, regulates testosterone production in testis. Following its binding to a G-couple receptor (Gprc6a) expressed in Leydig cells of the testes, osteocalcin promotes testosterone production by testis in a cAMP response element binding (CREB)-dependent manner. Testosterone is a sex steroid hormone required for many aspects of testicular functions, including germ cell survival and spermatogenesis. In addition to its role in reproductive function, testosterone actively regulates bone physiology. In particular, testosterone is a crucial determinant of bone growth during puberty, maturation, and maintenance of bone mass accrual.

phenotypes similar to those seen in $Ocn^{-/-}$ mice.[61,62] Furthermore, it had been proposed that Gprc6a was a calcium sensing receptor working better in the presence of osteocalcin.[61] Several criteria identified formally Gprc6a as an osteocalcin receptor present in Leydig cells. First, there is direct binding of osteocalcin to WT but not to *Gprc6a*-deficient Leydig cells; second, osteocalcin increases cAMP production in WT but not in *Gprc6a*-deficient Leydig cells; third, and more to the point, a Leydig cell-specific deletion of *Gprc6a* revealed a reproduction phenotype caused by low testosterone production similar, if not identical, to the one seen in the case of osteocalcin inactivation; fourth, in an even more convincing experiment, compound heterozygous mice lacking one copy of *Ocn* and one copy of *Gprc6a* had a reproduction phenotype identical in all aspects to the one seen in $Ocn^{-/-}$ or $Gprc6a^{-/-}$ mice.[8] The identification of Gprc6a as an osteocalcin receptor led to the demonstration that CREB is a transcriptional effector of osteocalcin regulation of testosterone biosynthesis by favoring the expression of key enzymes of this biosynthetic pathway in Leydig cells (Fig. 2). Interestingly, it does not affect the expression of the aromatase, responsible for estro-

gen synthesis.[8] The identification of an osteocalcin receptor now allows us to address many questions, for instance, to search for additional functions of osteocalcin. It also allows us to perform a more sophisticated dissection of the osteocalcin molecular mode of action in known, and yet to be identified, target cells.

Lately, the function of osteocalcin as a regulator of testosterone production has been extended to men. Dr. Khosla and colleagues showed recently that there is a significant association between serum osteocalcin and testosterone levels during mid-puberty in males, but not in females.[63] They postulate that this axis may be most relevant during rapid skeletal growth in adolescent males to help maximize bone size.

Conclusions

Beyond bone biology, this work demonstrates how intimately connected are all the organs and the power of genetic manipulations in model organisms. When it comes to bone and gonad biology, these findings reveal a previously unknown regulation of a gonad-derived hormone, testosterone. An obvious question raised by this novel role of osteocalcin is whether the skeleton also regulates some aspects of the reproductive function in humans. This notion is even more interesting since a recent study showed that the *Gprc6a* gene is related to prostate cancer progression in humans.[64,65] This is consistent with the fact that *Ocn*$^{-/-}$ and *Gprc6a*$^{-/-}$ mice show a significant change in the ratio testosterone/estradiol, which is a crucial determinant in the development of benign and malignant prostate tumors associated with aging in men. Interestingly, it is well known that testosterone levels in men decrease with age. This decrease is always associated with an increase of estrogen and luteinizing hormone, which cannot increase testosterone production. There are also associated increases in body fat mass and decrease in sperm counts and sexual function. All of these phenotypes are also present in *Ocn*$^{-/-}$ mice. This observation would suggest that osteocalcin could be an antiaging hormone. However, the recent study showing a correlation between the testosterone peak and osteocalcin level at adolescence in men, but not in females, indicates that osteocalcin may also be involved earlier during rapid skeletal growth in pubertal males.[63] In other words, this correlative data in humans suggests that osteocalcin may act as a novel determinant of the skeletal sexual dimorphism at puberty.

The functions of the skeleton are not all known yet, but the ones we now know indicate that skeleton physiology affects many more organs and functions than the skeleton itself. These novel functions of bone suggest that the skeleton is an important member of an endocrine network affecting multiple functions in the body.

Conflicts of interest

The author declares no conflicts of interest.

References

1. Khosla, S. *et al.* 2001. Relationship of serum sex steroid levels to longitudinal changes in bone density in young versus elderly men. *J. Clin. Endocrinol. Metab.* **86:** 3555–3561.
2. Riggs, B.L. *et al.* 1998. Long-term effects of calcium supplementation on serum parathyroid hormone level, bone turnover, and bone loss in elderly women. *J. Bone Miner. Res.* **13:** 168–174.
3. Nakamura, T. *et al.* 2007. Estrogen prevents bone loss via estrogen receptor alpha and induction of Fas ligand in osteoclasts. *Cell* **130:** 811–823.
4. Riggs, B.L. *et al.* 2002. Sex steroids and the construction and conservation of the adult skeleton. *Endocr. Rev.* **23:** 279–302.
5. Khosla, S. & B.L. Riggs. 2005. Pathophysiology of age-related bone loss and osteoporosis. *Endocrinol. Metab. Clin. North Am.* **34:** 1015–1030, xi.
6. Vanderschueren, D. *et al.* 2004. Androgens and bone. *Endocr. Rev.* **25:** 389–425.
7. Lee, N.K. *et al.* 2007. Endocrine regulation of energy metabolism by the skeleton. *Cell* **130:** 456–469.
8. Oury, F. *et al.* 2010. Endocrine regulation of male fertility by the skeleton. *Cell* **144:** 796–809.
9. Harada, S. & G.A. Rodan. 2003. Control of osteoblast function and regulation of bone mass. *Nature* **423:** 349–355.
10. Rodan, G.A. & T.J. Martin. 2000. Therapeutic approaches to bone diseases. *Science* **289:** 1508–1514.
11. Teitelbaum, S.L. 2000. Osteoclasts, integrins, and osteoporosis. *J. Bone Miner. Metab.* **18:** 344–349.
12. Karsenty, G. 2006. Convergence between bone and energy homeostases: leptin regulation of bone mass. *Cell Metab.* **4:** 341–348.
13. Seeman, E. 2001. Clinical review 137: sexual dimorphism in skeletal size, density, and strength. *J. Clin. Endocrinol. Metab.* **86:** 4576–4584.
14. Kirmani, S. *et al.* 2009. Bone structure at the distal radius during adolescent growth. *J. Bone Miner. Res.* **24:** 1033–1042.
15. Venken, K. *et al.* 2006. Relative impact of androgen and estrogen receptor activation in the effects of androgens on trabecular and cortical bone in growing male mice: a study in the androgen receptor knockout mouse model. *J. Bone Miner. Res.* **21:** 576–585.
16. Wakley, G.K. *et al.* 1991. Androgen treatment prevents loss of cancellous bone in the orchidectomized rat. *J. Bone Miner. Res.* **6:** 325–330.
17. Turner, R.T. *et al.* 1990. Differential effects of androgens on cortical bone histomorphometry in gonadectomized male and female rats. *J. Orthop. Res.* **8:** 612–617.
18. Seeman, E. 2002. Pathogenesis of bone fragility in women and men. *Lancet* **359:** 1841–1850.
19. Katznelson, L. *et al.* 1996. Increase in bone density and lean body mass during testosterone administration in men with acquired hypogonadism. *J. Clin. Endocrinol. Metab.* **81:** 4358–4365.
20. Bertelloni, S. *et al.* 1995. Short-term effect of testosterone treatment on reduced bone density in boys with constitutional delay of puberty. *J. Bone Miner. Res.* **10:** 1488–1495
21. Finkelstein, J.S. *et al.* 1996. A longitudinal evaluation of bone mineral density in adult men with histories of delayed puberty. *J. Clin. Endocrinol. Metab.* **81:** 1152–1155.
22. Buchanan, J.R. *et al.* 1988. Determinants of peak trabecular bone density in women: the role of androgens, estrogen, and exercise. *J. Bone Miner. Res.* **3:** 673–680.
23. Zborowski, J.V. *et al.* 2001. Clinical Review 116: bone mineral density, androgens, and the polycystic ovary. The complex and controversial issue of androgenic influence in female bone. *J. Clin. Endocrinol. Metab.* **85:** 3496–3506.
24. Wei, S. *et al.* 2010. Menstrual irregularity and bone mass in premenopausal women: cross-sectional associations with testosterone and SHBG. *BMC Musculoskelet Disord.* **11:** 288–296.
25. Callewaert, F. *et al.* 2009. Skeletal sexual dimorphism: relative contribution of sex steroids, GH-IGF1, and mechanical loading. *J. Endocrinol.* **207:** 127–134.

26. Lupu, F. *et al.* 2001. Roles of growth hormone and insulin-like growth factor 1 in mouse postnatal growth. *Dev. Biol.* **229:** 141–162.

27. Callewaert, F. *et al.* 2010. Sex steroids and the male skeleton: a tale of two hormones. *Trends Endocrinol. Metab.* **21:** 89–95.

28. Stepan, J.J. *et al.* 1989. Castrated men exhibit bone loss: effect of calcitonin treatment on biochemical indices of bone remodeling. *J. Clin. Endocrinol. Metab.* **69:** 523–527.

29. Kaufman, J.M. & A. Vermeulen. 2005. The decline of androgen levels in elderly men and its clinical and therapeutic implications. *Endocr. Rev.* **26:** 833–876.

30. Araujo, A.B. *et al.* 2008. Correlations between serum testosterone, estradiol, and sex hormone-binding globulin and bone mineral density in a diverse sample of men. *J. Clin. Endocrinol. Metab.* **93:** 2135–2141.

31. Mellstrom, D. *et al.* 2006. Free testosterone is an independent predictor of BMD and prevalent fractures in elderly men: MrOS Sweden. *J. Bone Miner. Res.* **21:** 529–535.

32. van den Beld, A.W. *et al.* 2000. Measures of bioavailable serum testosterone and estradiol and their relationships with muscle strength, bone density, and body composition in elderly men. *J. Clin. Endocrinol. Metab.* **85:** 3276–3282.

33. Bouillon, R. *et al.* 2004. Estrogens are essential for male pubertal periosteal bone expansion. *J. Clin. Endocrinol. Metab.* **89:** 6025–6029.

34. Rochira, V. *et al.* 2007. Skeletal effects of long-term estrogen and testosterone replacement treatment in a man with congenital aromatase deficiency: evidences of a priming effect of estrogen for sex steroids action on bone. *Bone* **40:** 1662–1668.

35. Imai, Y. *et al.* 2009. Estrogens maintain bone mass by regulating expression of genes controlling function and life span in mature osteoclasts. *Ann. N.Y. Acad. Sci.* **1173**(Suppl 1): E31–E39.

36. Hughes, D.E. *et al.* 1996. Estrogen promotes apoptosis of murine osteoclasts mediated by TGF-beta. *Nat. Med.* **2:** 1132–1136.

37. Majeska, R.J. *et al.* 1994. Direct modulation of osteoblastic activity with estrogen. *J. Bone Joint Surg. Am.* **76:** 713–721.

38. Qu, Q. *et al.* 1998. Estrogen enhances differentiation of osteoblasts in mouse bone marrow culture. *Bone* **22:** 201–209.

39. Xing, L. & B.F. Boyce. 2005. Regulation of apoptosis in osteoclasts and osteoblastic cells. *Biochem. Biophys. Res. Commun.* **328:** 709–720.

40. Jimi, E. *et al.* 1996. Interleukin-1 alpha activates an NF-kappaB-like factor in osteoclast-like cells. *J. Biol. Chem.* **271:** 4605–4608.

41. Pacifici, R. 1996. Estrogen, cytokines, and pathogenesis of postmenopausal osteoporosis. *J. Bone Miner. Res.* **11:** 1043–1051.

42. Manolagas, S.C. *et al.* 2002. Sex steroids and bone. *Recent Prog. Horm. Res.* **57:** 385–409.

43. Tau, K.R. *et al.* 1998. Estrogen regulation of a transforming growth factor-beta inducible early gene that inhibits deoxyribonucleic acid synthesis in human osteoblasts. *Endocrinology* **139:** 1346–1353.

44. Gill, R.K. *et al.* 1998. Orchiectomy markedly reduces the concentration of the three isoforms of transforming growth factor beta in rat bone, and reduction is prevented by testosterone. *Endocrinology* **139:** 546–550.

45. Bodine, P.V. *et al.* 1995. Regulation of c-fos expression and TGF-beta production by gonadal and adrenal androgens in normal human osteoblastic cells. *J. Steroid Biochem. Mol. Biol.* **52:** 149–158.

46. Michael, H. *et al.* 2005. Estrogen and testosterone use different cellular pathways to inhibit osteoclastogenesis and bone resorption. *J. Bone Miner. Res.* **20:** 2224–2232.

47. Krum, S.A. *et al.* (2008) Estrogen protects bone by inducing Fas ligand in osteoblasts to regulate osteoclast survival. *EMBO J.* **27:** 535–545.

48. Jilka, R.L. *et al.* 1992. Increased osteoclast development after estrogen loss: mediation by interleukin-6. *Science* **257:** 88–91.

49. Kasperk, C.H. *et al.* 1989. Androgens directly stimulate proliferation of bone cells *in vitro. Endocrinology* **124:** 1576–1578.

50. Kousteni, S. *et al.* 2002. Reversal of bone loss in mice by nongenotropic signaling of sex steroids. *Science* **298:** 843–846.

51. Kousteni, S. *et al.* 2001. Nongenotropic, sex-nonspecific signaling through the estrogen or androgen receptors: dissociation from transcriptional activity. *Cell* **104:** 719–730.

52. Noble, B. *et al.* 1999. Androgen receptors in bone-forming tissue. *Horm. Res.* **51:** 31–36.

53. Bord, S. *et al.* 2001. Estrogen receptors alpha and beta are differentially expressed in developing human bone. *J. Clin. Endocrinol. Metab.* **86:** 2309–2314.

54. Couse, J.F. & K.S. Korach. 1999. Estrogen receptor null mice: what have we learned and where will they lead us? *Endocr. Rev.* **20:** 358–417.

55. Vidal, O. *et al.* 2000. Estrogen receptor specificity in the regulation of skeletal growth and maturation in male mice. *Proc. Natl. Acad. Sci. USA* **97:** 5474–5479.

56. Oz, O.K. *et al.* 2000. Bone has a sexually dimorphic response to aromatase deficiency. *J. Bone Miner. Res.* **15:** 507–514.

57. Lindberg, M.K. *et al.* 2001 Estrogen receptor specificity in the regulation of the skeleton in female mice. *J. Endocrinol.* **171:** 229–236.

58. Ferron, M. *et al.* 2010. Insulin signaling in osteoblasts integrates bone remodeling and energy metabolism. *Cell* **142:** 296–308.

59. Fulzele, K. *et al.* 2010. Insulin receptor signaling in osteoblasts regulates postnatal bone acquisition and body composition. *Cell* **142:** 309–319.

60. Yoshikawa, Y. *et al.* 2011. Genetic evidence points to an osteocalcin-independent influence of osteoblasts on energy metabolism. *J. Bone Miner. Res.* **26:** 2012–2025.

61. Pi, M. *et al.* 2008. GPRC6A null mice exhibit osteopenia, feminization and metabolic syndrome. *PLoS One* **3:** e3858.

62. Pi, M. *et al.* 2011. GPRC6A mediates responses to osteocalcin in beta-cells in vitro and pancreas in vivo. *J. Bone Miner. Res.* **26:** 1680–1683.

63. Kirmani, S. *et al.* 2011. Relationship of testosterone and osteocalcin levels during growth. *J. Bone Miner. Res.* **26:** 2212–2216.

64. Takata, R. *et al.* 2010. Genome-wide association study identifies five new susceptibility loci for prostate cancer in the Japanese population. *Nat. Genet.* **42:** 751–754.

65. Pi, M. & L.D. Quarles. GPRC6A regulates prostate cancer progression. *Prostate* **21442:** 1–11.

Ann. N.Y. Acad. Sci. ISSN 0077-8923

ANNALS OF THE NEW YORK ACADEMY OF SCIENCES
Issue: *Blavatnik Awards for Young Scientists*

Calabi-Yau manifolds and their degenerations

Valentino Tosatti

Department of Mathematics, Columbia University, New York, New York

Address for correspondence: Valentino Tosatti, Department of Mathematics, Columbia University, Room 625 MC 4444, 2990 Broadway, New York, NY 10027. tosatti@math.columbia.edu

Calabi-Yau manifolds are geometric objects of central importance in several branches of mathematics, including differential geometry, algebraic geometry, and mathematical physics. In this paper, we give a brief introduction to the subject aimed at a general mathematical audience and present some of our results that shed some light on the possible ways in which families of Calabi-Yau manifolds can degenerate.

Keywords: Calabi-Yau manifolds; Kähler metrics; Ricci curvature; degenerations of manifolds

Introduction

While preparing this paper we faced a difficult choice: should the paper be addressed to a general science audience, or should it be written primarily for scholars in our field? We decided to strike a balance between these two approaches and write a paper that could be read by any mathematician. We apologize to all the other potential readers for this choice. The interested reader can find a gentle introduction to this topic in the recent book of Yau and Nadis[1] as well as in their article.[2]

Definitions of Calabi-Yau manifolds

The main objects of study in this paper are Calabi-Yau manifolds. There are many possible definitions of these spaces, and we will start by reviewing a few of them. First of all, recall that a complex manifold X is a smooth manifold of real dimension $2n$ with an atlas whose transition functions are holomorphic. In particular, each tangent space of X is naturally identified with C^n, and multiplication by i induces a tensor $J: TX \rightarrow TX$ with $J^2 = -Id$. A Riemannian metric g on X is called Hermitian if $g(JY,JZ) = g(Y,Z)$ for all vectors Y,Z. In this case, we define $\omega(Y,Z) = g(JY,Z)$, which is a skew-symmetric 2-form on X. If $d\omega = 0$, we say that g is a Kähler metric, and we say that X is Kähler if it admits Kähler metrics. The cohomology class $[\omega]$ lives in $H^2(X,R) \cap H^{1,1}(X) =: H^{1,1}(X,R)$ and is called a Kähler class. As we just explained, the tangent bundle of a complex mani-

fold inherits a complex structure J, and so it has well-defined Chern classes $c_i(X)$ in $H_i^2(X,Z)$, $1 \leq i \leq n$. We can now give several equivalent definitions of Calabi-Yau manifolds:

(1) complex geometry: a Calabi-Yau manifold is a compact Kähler manifold X with first Chern class $c_1(X)$ equal to zero in the cohomology group $H^2(X,R)$;
(2) algebraic geometry: a Calabi-Yau manifold is a compact Kähler manifold X with torsion canonical bundle $K = \Lambda^n T^{1,0} X^*$, so that $mK = O_X$ for some integer $m \geq 1$;
(3) Einstein equation: a Calabi-Yau manifold is a compact complex manifold X with a Kähler metric ω with Ricci curvature identically zero (Ricci-flat); and
(4) Riemannian geometry: a Calabi-Yau manifold is a compact Riemannian manifold (X,g) of real dimension $2n$ with restricted holonomy group contained in the special unitary group $SU(n)$.

We will explain why these definitions are equivalent after giving a few examples.

Examples of Calabi-Yau manifolds

The following are some simple examples of Calabi-Yau manifolds.

Example 1. Let $X = C^n/\Lambda$ be the quotient of Euclidean space C^n by a lattice $\Lambda = Z^{2n}$. Then

doi: 10.1111/j.1749-6632.2011.06259.x

Ann. N.Y. Acad. Sci. 1260 (2012) 8–13 © 2012 New York Academy of Sciences.

X is topologically a torus $(S^1)^{2n}$ and it has trivial tangent bundle and therefore also trivial canonical bundle. All Calabi-Yau manifolds of complex dimension $n = 1$ are tori and are also called *elliptic curves*.

Example 2. A Calabi-Yau manifold with complex dimension $n = 2$, which is also simply connected, is called a *K3 surface*. Every Calabi-Yau surface is either a torus, a *K3* surface, or a finite unramified quotient of these. In general, these quotients will have torsion but nontrivial canonical bundle, as is the case, for example, for *Enriques surfaces*, which are Z_2 quotients of *K3*.

Example 3. Let X be a smooth complex hypersurface of degree $n + 2$ inside complex projective space CP^{n+1}. Then by the adjuction formula the canonical bundle of X is trivial, and so X is a Calabi-Yau manifold. When $n = 1$ we get an elliptic curve, and when $n = 2$ a *K3* surface. More generally, one can consider smooth complete intersections in product of projective spaces, with suitable degrees, and get more examples of Calabi-Yau manifolds.

Example 4. Let $T = C^2/\Lambda$ be a torus of complex dimension 2 and consider the reflection through the origin $i:C^2 \rightarrow C^2$. This descends to an involution of T with 16 fixed points, and we can take the quotient $Y = T/i$, which is an algebraic variety with 16 singular rational double points (also known as *orbifold* points). We resolve these 16 points by blowing them all up and we get a map $f:X \rightarrow Y$ where X is a smooth *K3* surface, known as the *Kummer surface* of the torus T.

Example 5. A Calabi-Yau manifold (X,g) of even complex dimension n with holonomy equal to the symplectic group $Sp(n/2) \subset SU(n)$ is called *hyperkähler*. We have that $Sp(1) = SU(2)$ so the only hyperkähler manifolds of complex dimension 2 are *K3* surfaces. There are not many examples of higher-dimensional hyperkähler manifolds. The simplest one is obtained by taking a *K3* surface Y and looking at $\hat{Y} \rightarrow Y \times Y$, the blow-up of the diagonal. Flipping the two factors of $Y \times Y$ induces a $Z/2$-action on $\hat{Y}$, and the quotient space X is hyperkähler.

Let us now discuss why the four different definitions of Calabi-Yau manifolds that we gave are equivalent. The fact that definitions (3) and (4) are equivalent is a simple exercise in Riemannian geometry, and to see that (2) implies (1), it suffices to take the first Chern class of the canonical bundle.

The implication from (1) to (3) is the content of the celebrated Calabi Conjecture:[3]

Theorem 1. (Yau's solution of the Calabi Conjecture, 1976, see Refs. 4,5.) On any compact Kähler manifold X with $c_1(X) = 0$ in $H^2(X,R)$, there exist Kähler metrics with Ricci curvature identically zero. Moreover, there is a unique such Ricci-flat metric in each Kähler class of X.

Finally, the fact that (3) implies (2) is a consequence of a decomposition theorem due to Yau:[3,6] every compact Ricci-flat Kähler manifold of complex dimension n has a finite unramified cover $\tilde{X}$ that splits isometrically as a product of a flat torus, of simply connected Calabi-Yau manifolds with holonomy equal to $SU(n)$, and of simply connected hyperkähler manifolds. In particular, the canonical bundle of $\tilde{X}$ is trivial, which implies that the canonical bundle of X is torsion.

Let us now briefly discuss why Calabi-Yau manifolds are important in several branches of mathematics. They are important in differential geometry because they give examples of Einstein metrics, which are Ricci-flat but not flat if X is not covered by a torus, and these metrics are almost never explicit. Furthermore, if the Kähler class is fixed, the Ricci-flat metric is uniquely determined by the complex structure. Using this fact, Tian[7] and Todorov[8] proved that the moduli space of polarized Calabi-Yau manifolds is smooth.

They are important in algebraic geometry because of the position they occupy in the theory of classification of algebraic varieties. These are usually divided into families according to their Kodaira dimension $\kappa(x) \in \{-\infty, 0, 1, \ldots, n\}$, where $n = \dim_C(X)$, and conjecturally, every algebraic variety with $\kappa(X) = 0$ is birational to a Calabi-Yau variety (possibly singular). There are many other basic questions about Calabi-Yau manifolds that are still open: does every simply connected Calabi-Yau manifold have a rational curve? Is the number of deformation classes of simply connected Calabi-Yau manifolds of a given dimension finite?

Calabi-Yau threefolds with holonomy $SU(3)$ are important in mathematical physics because they can be used in string theory to construct supersymmetric theories that live (locally) on the product of four-dimensional spacetime with a Calabi-Yau threefold. This led to the discovery of the mathematical phenomenon of mirror symmetry, which has generated a huge amount of research.[9] In the general framework of mirror symmetry, one has a family of Calabi-Yau manifolds parametrized by a punctured disc Δ^* in the complex plane, which degenerates when approaching the origin. The chief example of this is the family of Calabi-Yau quintic hypersurfaces X_t in CP^4 given by the equation

$$Z_0^5 + Z_1^5 + Z_2^5 + Z_3^5 + Z_4^5 = 5t^{-1} Z_0 Z_1 Z_2 Z_3 Z_4$$

in homogeneous coordinates. As t approaches zero, X_t degenerates to a union of five hyperplanes. This is an example of a *large complex structure limit*. To such a family, one then associates a "mirror family" of Calabi-Yau manifolds $\hat{X}_t$ with fixed complex structure and varying Kähler class, which approaches a *large Kähler structure limit*. Then for t close to zero, one can relate invariants of the complex structure of X_t to invariants of the symplectic structure of $\hat{X}_t$. For an introduction to this circle of ideas, see the book by Gross, Huybrechts, and Joyce.[10]

Finally, let us note here that there are many possible generalizations of the concept of Calabi-Yau manifolds: noncompact ones, singular ones, non-Kähler ones, symplectic ones, and so on. We will not delve here into all these concepts and their precise definitions, but refer the reader to the survey article of Yau[11] instead.

Degenerations of Calabi-Yau manifolds

From the discussion in the previous section, it is apparent that the study of degenerations of families of Calabi-Yau manifolds is an interesting topic with ramifications in several different branches of mathematics. In the rest of this paper, we will study the behavior of Ricci-flat Kähler metrics on a fixed Calabi-Yau manifold as their Kähler class degenerates. If the complex structure is allowed to vary, similar results were obtained by Ruan and Zhang[12] and Rong and Zhang[13] building upon our work.

First of all, let us identify the parameter space for Kähler classes on a compact Kähler manifold X. Recall that a Kähler class on X is a cohomology class α in $H^{1,1}(X,R)$ that can be written as $\alpha = [\omega]$ for some Kähler metric ω. The set of all Kähler classes is called the *Kähler cone* of X and is an open convex cone $K_X \subset H^{1,1}(X,R)$ that has the origin as its vertex.

Question 1. What is the behavior of these Ricci-flat Kähler metrics when the class α degenerates to the boundary of the Kähler cone?

This question was posed by many people, including Yau,[14,15] Wilson,[16] and McMullen.[17] To get a feeling for what the Kähler cone and its boundary represent geometrically, we start with the following observation. If $V \subset X$ is a complex subvariety of complex dimension $k > 0$, then it is well known (from the work of Lelong) that V defines a homology class $[V]$ in $H_{2k}(X,Z)$. Moreover, if $[\omega]$ is a Kähler class, the pairing $<[V],[\omega]^k>$ equals $\int_V \omega^k = k!Vol(V,\omega) > 0$, the volume of V with respect to the Kähler metric ω (Wirtinger's Theorem). It follows that if a class α is on the boundary of K_X and if V is any complex subvariety, then the pairing $<[V],[\alpha]^k>$ is nonnegative, and moreover, a theorem of Demailly-Paun[18] shows that there must be subvarieties V with pairing zero. Therefore, as we approach the class α from inside K_X, these subvarieties have volume that goes to zero, and the Ricci-flat metrics must degenerate (in some way) along these subvarieties.

We now make Question 1 more precise. On a compact Calabi-Yau manifold X fix a nonzero class α_0 on the boundary of K_X and let $\{\alpha_t\}_{0 \leq t \leq 1}$ be a smooth path of classes in $H^{1,1}(X,R)$ originating at α_0 and with $\alpha_t \in K_X$ for $t > 0$. Call ω_t the unique Ricci-flat Kähler metric on X cohomologous to α_t for $t > 0$, which is produced by Theorem 1.

Question 2. What is the behavior of the Ricci-flat metrics ω_t when t goes to zero?

Of course, we could also consider sequences of classes instead of a path, and all we are going to say works equally well in that case. Notice that we are not allowing the class α_t to go to infinity in $H^{1,1}(X,R)$ as it approaches ∂K_X. Because of this, we can prove the following basic fact, independently discovered by Zhang:[12]

Theorem 2. (Tosatti,[19] Zhang[12]) The diameter of the metrics ω_t has a uniform upper bound as t approaches zero, $diam(X,\omega_t) < C$.

On the other hand, it is easy to construct examples of Ricci-flat Kähler metrics with unbounded cohomology class that violate Theorem 2, just by rescaling a fixed metric by a large number. Going back to Question 2, the problem splits naturally into two cases that exhibit a rather different behavior, according to whether the total integral $\int_X \alpha_0^n$ is strictly positive or zero. If $\int_X \alpha_0^n$ is positive, this means that the volume $n!Vol(X,\omega_t) = \int_X \omega_t^n = \int_X \alpha_t^n$ remains bounded away from zero as t goes to 0, and this is called the *noncollapsing* case. If $\int_X \alpha_0{}^n = 0$, then the volume $Vol(X, \omega_t)$ converges to zero, and this is called the *collapsing* case.

The main Question 2 falls into the general problem of understanding limits of sequences of Einstein manifolds with an upper bound for the diameter (but no bound for the sectional curvature in general), a topic that has been extensively studied (e.g., see Refs. 20–23). Our results are of a quite different nature from these works and give stronger conclusions. The first theorem gives a satisfactory answer in the noncollapsing case.

Theorem 3. (Tosatti[19]) If $\int_X \alpha_0{}^n > 0$ then the Ricci-flat metrics ω_t converge smoothly away from an analytic subvariety S to an incomplete Ricci-flat metric on its complement.

In fact, the subvariety S in this theorem is simply the union of all complex subvarieties where α_0 integrates to zero. Whenever α_0 is a rational class, the limit can also be understood using algebraic geometry: the subvariety S can be contracted to create a singular Calabi-Yau manifold, and the limit metric is the pullback of a Ricci-flat metric on the smooth part.

In the collapsing case, when $\int_X \alpha_0{}^n = 0$, things are more complicated. If X is projective and the class α_0 is rational, then the Log Abundance Conjecture in algebraic geometry implies that there is a fibration $f: X \to Y$ where Y is an algebraic variety of strictly lower dimension m and α_0 is the pullback of an ample class on Y. If we call S the critical locus of f inside X, then S is a subvariety and $f: X \setminus S \to Y \setminus f(S)$ is a smooth submersion with fibers Calabi-Yau manifolds $X_y = f^{-1}(y)$ of complex dimension $n{-}m$. The subvariety S is the union of all singular fibers of f together with all the fibers with dimensions strictly larger than $n{-}m$. We also take $\alpha_t = \alpha_0 + t[\omega_X]$ where ω_X is a fixed Kähler metric on X. We then have the following result, which says that the Ricci-flat metrics shrink the manifold to the base of the fibration.

Theorem 4. (Tosatti[24]) Let $f: X \to Y$ be such a holomorphic fibration and $\alpha_t = \alpha_0 + t[\omega_X]$. Then there is a smooth Kähler metric ω on $Y \setminus f(S)$ such that when t approaches zero the Ricci-flat metrics ω_t converge to $f^*\omega$. The metric ω has Ricci curvature equal to a Weil–Petersson metric that measures the change of complex structures of the Calabi-Yau fibers X_y.

If we furthermore assume that X is projective and all the smooth fibers X_y are tori, then we can prove a stronger result:

Theorem 5. (Gross *et al.*[25]) In the same setting as Theorem 4, assume that X is projective and the smooth fibers X_y are tori. Then the convergence of ω_t to $f^*\omega$ is smooth and the sectional curvature of ω_t remains locally bounded on $X \setminus S$. Along each torus fiber X_y the rescaled metrics $t^{-1}\omega_t|_{X_y}$ converge smoothly to a flat metric. Finally, for any Gromov–Hausdorff limit space (Z, d) of (X, ω_t), there is a local isometric embedding $(Y \setminus S, \omega) \to (Z, d)$ with dense image.

Theorem 5 can also be applied to study the large complex structure limits of families of polarized hyperkähler manifolds in the large complex structure limit.[25]

Examples of degenerations

First of all, notice that Question 2 is only interesting if $\dim H^{1,1}(X, R) > 1$, because otherwise K_X reduces to an open half-line and there is only one Ricci-flat Kähler metric on X up to global scaling by a constant, so the only possible degenerations are given by scaling this metric to zero or infinity. For this reason, Question 2 is essentially void on Calabi-Yau manifolds of dimension $n = 1$ (i.e., elliptic curves).

Example 6. Let $X = C^n/\Lambda$ be a complex torus. A Ricci-flat Kähler metric on X is the same as a flat Kähler metric, and each flat metric can be identified simply with a positive definite Hermitian $n \times n$ matrix. The boundary of the Kähler cone is then represented by nonnegative definite Hermitian matrices H with nontrivial kernel $\Sigma \subset C^n$ (notice that in this case every class on ∂K_X has zero integral, so we are always in the collapsing case).

If the class α_0 corresponds to such a matrix H with the kernel Σ, which is Q-defined modulo Λ, then we can quotient Σ out and get a map $f: X \to Y = C^m/\Lambda'$ to a lower-dimensional torus ($m < n$) such that $H = f^*H'$ with H' a positive definite $m \times m$ Hermitian matrix. It follows that when t approaches zero, the (Ricci-) flat metrics ω_t collapse to the flat metric on Y that corresponds to H'. This is a special case of Theorems 4 and 5, with S empty and Weil–Petersson metric identically zero (since all the torus fibers are isomorphic).

If on the other hand the kernel Σ is not Q-defined, then Σ defines a foliation on X (which is not a fibration anymore) and the limit H of the (Ricci-) flat metrics is a smooth nonnegative form that is *transversal* to the foliation (that is, positive in the complementary directions). This case gives an idea of what to expect in general in the collapsing case when there is no fibration.

Example 7. Let $f{:}X{\rightarrow}Y$ be the Kummer $K3$ surface of a torus T, where $Y = T/i$ is the singular quotient of T and f is the blowup map. Take α_0 to be the pullback of an ample divisor on Y, and note that $\int_X \alpha_0^2 > 0$. If we call S the union of the 16 exceptional divisors of f—that is, the union of the 16 spheres S^2, which are the preimages of the singular points of Y—then S is a complex submanifold of X. Then Kobayashi and Todorov[26] (using classical results on the moduli space of $K3$ surfaces, such as the Torelli theorem) proved that for any path α_t of Kähler classes that approach α_0, the Ricci-flat metrics ω_t converge smoothly away from S to the pullback of the unique flat orbifold metric on Y cohomologous to the ample divisor we chose. Here, an orbifold flat metric on Y simply means a flat metric on T, which is invariant under i. This statement is a special case of our Theorem 3, which in particular gives a new proof of the result of Kobayashi and Todorov.

Example 8. Let X be a $K3$ surface that admits an elliptic fibration $f{:}X{\rightarrow}Y = CP^1$. This means that f is a surjective holomorphic map with all the fibers smooth elliptic curves except a finite number of fibers S, which are singular elliptic curves. Again, we take α_0 to be the pullback of an ample divisor on Y and note that $\int_X \alpha_0^2 = 0$. We also take $\alpha_t = \alpha_0 + t[\omega_X]$ for a Kähler metric ω_X. Then Gross and Wilson[27] have shown that when t goes to zero, the metrics ω_t converge smoothly away from S to the pullback $f^*\omega$, where ω is a Kähler metric on $Y = CP^1$ minus the finitely many points $f(S)$ with singular preimage. Moreover, they showed that away from S, the rescaled Ricci-flat metrics along the fibers $t^{-1}\omega_t|_{Xy}$ converge to flat metrics. More recently, Song and Tian[28] have noticed that the metric ω on $CP^1 \setminus f(S)$ has Ricci curvature equal to the pullback of the Weil–Petersson metric from the moduli space of elliptic curves via the map that to a point in $CP^1 \setminus f(S)$ associates the elliptic curve that lies above that point. This result is a special case of our Theorems 4 and 5, which also provide a new proof of the theorem of Gross and Wilson.

Questions

Let us now mention a few open problems related to the above results, which we find very interesting.

Question 3. In the same setting as Theorem 4 prove that the rescaled Ricci-flat metrics along the fibers $t^{-1}\omega_t|_{Xy}$ converge smoothly to the unique Ricci-flat Kähler metric on X_y cohomologous to $[\omega_X]|_{Xy}$.

As explained in our work,[25] this would be implied by the following:

Question 4. In the same setting as Theorem 4 prove that the convergence of ω_t to $f^*\omega$ is smooth away from S.

Both Questions 3 and 4 are solved in our work[25] when X is projective and the smooth fibers are tori.

The natural remaining question is what happens to the Ricci-flat metrics ω_t when α_0 is an irrational class with $\int_X \alpha_0^n = 0$, so that there is no fibration structure. We conjecture the following:

Question 5. In this situation there is a subvariety $S \subset X$ and a smooth nonnegative $(1,1)$ form ω_0 on $X \setminus S$, which satisfies $\omega_0^n = 0$, so that the Ricci-flat metrics ω_t converge smoothly away from S to ω_0.

In this case, taking the kernel of ω_0 we would get a foliation on $X \setminus S$ with leaves holomorphic subvarieties, a so-called "Monge-Ampère foliation." Of course, Question 5 is correct in the case of tori, as in Example 6.

Another problem that seems very interesting is whether the convergence in Theorems 3 and 4 holds in the Gromov–Hausdorff sense. More precisely, in Theorem 3 consider the metric space completion (Z,d) of $(X \setminus S,\omega_0)$, while in Theorem 4 call (Z,d) the metric space completion of $(Y \setminus f(S),\omega)$.

Question 6. In the setting of either Theorem 3 or 4, do the Ricci-flat manifolds (X,ω_t) converge to (Z,d) in the Gromov–Hausdorff sense? Moreover, in Theorem 4, is Z homeomorphic to Y the algebrogeometric limit?

The Gromov–Hausdorff convergence is proved for $K3$ surfaces in the noncollapsing case in our work,[19] and there are further results in the noncollapsing case by Ruan and Zhang.[12] In the case of collapsing $K3$ surfaces, this was proved by Gross and Wilson.[27] Finally, we pose the following:

Question 7. Let X be a compact Calabi-Yau manifold and α a cohomology class on ∂K_X. Does there exist a smooth nonnegative form ω cohomologous to α?

If X is projective and α is rational, then this can be proved using algebraic geometry if $\int_X \alpha^n > 0$, and it follows from the Log Abundance Conjecture if $\int_X \alpha^n = 0$, see our work.[19]

Acknowledgment

This work was partially supported by National Science Foundation Grant DMS-1005457.

Conflicts of interest

The author declares no conflicts of interest.

References

1. Yau, S.-T. & S. Nadis. 2010. *The Shape of Inner Space. String Theory and the Geometry of the Universe's Hidden Dimensions*. Basic Books. New York.
2. Yau, S.-T. & S. Nadis. 2011. String theory and the geometry of the Universe's hidden dimensions. *Notices Amer. Math. Soc.* **58**: 1067–1076.
3. Calabi, E. 1957. On Kähler manifolds with vanishing canonical class. In *Algebraic Geometry and Topology. A Symposium in Honor of S. Lefschetz*. R.H. Fox, D.C. Spencer & A.W. Tucker, Eds.: 78–89. Princeton University Press. Princeton.
4. Yau, S.-T. 1977. Calabi's conjecture and some new results in algebraic geometry. *Proc. Nat. Acad. Sci. U.S.A.* **74**: 1798–1799.
5. Yau, S.-T. 1978. On the Ricci curvature of a compact Kähler manifold and the complex Monge-Ampère equation, I. *Comm. Pure Appl. Math.* **31**: 339–411.
6. Beauville, A. 1983. Variétés Kähleriennes dont la première classe de Chern est nulle. *J. Differential Geom.* **18**: 755–782.
7. Tian, G. 1987. *Smoothness of the Universal Deformation Space of Compact Calabi-Yau Manifolds and Its Petersson-Weil Metric. Mathematical Aspects of String Theory (San Diego, Calif., 1986)*, World Scientific Publishing. Singapore.
8. Todorov, A.N. 1989. The Weil-Petersson geometry of the moduli space of SU(n≥3) (Calabi-Yau) manifolds. *I. Comm. Math. Phys.* **126**: 325–346.
9. Strominger, A., S.-T. Yau & E. Zaslow. 1996. Mirror symmetry is T-duality. *Nuclear Phys. B* **479**: 243–259.
10. Gross, M., D. Huybrechts & D. Joyce. 2003. *Calabi-Yau Manifolds and Related Geometries*. Springer-Verlag. Berlin.
11. Yau, S.-T. 2009. A survey of Calabi-Yau manifolds. In *Surveys in Differential Geometry. Vol. XIII: Geometry, Analysis, and Algebraic Geometry: Forty Years of the Journal of Differential Geometry*. H.D. Cao & S.T. Yau, Eds.: 277–318. International Press. Somerville, MA.
12. Ruan W.-D. & Y. Zhang. 2011. Convergence of Calabi-Yau manifolds. *Adv. Math.* **228**: 1543–1589.
13. Rong, X. & Y. Zhang. 2010. Continuity of extremal transitions and flops for Calabi-Yau manifolds. arXiv:1012.2940.
14. Yau, S.-T. 1982. Problem section. In *Seminar on Differential Geometry*. S.T. Yau, Ed.: 669–706. Princeton University Press. Princeton.
15. Yau, S.-T. 1993. Open problems in geometry. *Proc. Symp. Pure Math.* **54**: 1–28.
16. Wilson, P.M.H. 2004. Metric limits of Calabi-Yau manifolds. In *The Fano Conference*. University of Torino. Turin.
17. McMullen, C.T. 2002. Dynamics on K3 surfaces: Salem numbers and Siegel disks. *J. Reine Angew. Math.* **545**: 201–233.
18. Demailly, J.-P. & M. Păun. 2004. Numerical characterization of the Kähler cone of a compact Kähler manifold. *Ann. Math.* **159**: 1247–1274.
19. Tosatti, V. 2009. Limits of Calabi-Yau metrics when the Kähler class degenerates. *J. Eur. Math. Soc. (JEMS).* **11**: 755–776.
20. Gromov, M. 1999. *Metric Structures for Riemannian and Non-Riemannian Spaces*. Birkähuser. Boston.
21. Anderson, M.T. 1989. Ricci curvature bounds and Einstein metrics on compact manifolds. *J. Amer. Math. Soc.* **2**: 455–490.
22. Bando, S., A. Kasue & H. Nakajima. 1989. On a construction of coordinates at infinity on manifolds with fast curvature decay and maximal volume growth. *Invent. Math.* **97**: 313–349.
23. Cheeger, J. & T. Colding. 1997. On the structure of spaces with Ricci curvature bounded below. *I. J. Different. Geom.* **46**: 406–480.
24. Tosatti, V. 2010. Adiabatic limits of Ricci-flat Kähler metrics. *J. Different. Geom.* **84**: 427–453.
25. Gross, M., V. Tosatti & Y. Zhang. 2011. Collapsing of Abelian fibred Calabi-Yau manifolds. arXiv:1108.0967.
26. Kobayashi, R. & A.N. Todorov. 1987. Polarized period map for generalized K3 surfaces and the moduli of Einstein metrics. *Tohoku Math. J.* **39**: 341–363.
27. Gross, M. & P.M.H. Wilson. 2000. Large complex structure limits of K3 surfaces. *J. Different. Geom.* **55**: 475–546.
28. Song, J. & G. Tian. 2007. The Kähler-Ricci flow on surfaces of positive Kodaira dimension. *Invent. Math.* **170**: 609–653.

Ann. N.Y. Acad. Sci. ISSN 0077-8923

ANNALS OF THE NEW YORK ACADEMY OF SCIENCES
Issue: *Blavatnik Awards for Young Scientists*

Emerging topics in epigenetics: ants, brains, and noncoding RNAs

Roberto Bonasio

Howard Hughes Medical Institute, Department of Biochemistry, New York University School of Medicine, New York, New York

Address for correspondence: Roberto Bonasio, Department of Biochemistry, NYU School of Medicine, 522 First Avenue, SML207, New York, NY 10016. roberto.bonasio@nyumc.org

One of the greatest wonders in biology is the high degree of molecular organization and complexity achieved by multicellular life forms, which are typically composed by hundreds of cell types, each with a unique identity and function and all sharing the same genome. Long-term maintenance of these distinct cell identities requires epigenetic signals, molecular signatures that regulate gene expression and can be inherited during cell division. Some epigenetic signals also appear to have an intimate connection with brain function, with important implications for neuroscience and medicine. To better understand these phenomena, new technologies must be developed and nonconventional model organisms should be studied. For example, the genomes of eusocial insects, such as ants and honeybees, specify drastically different morphologies (polyphenism) and behaviors (polyethism) that yield adult individuals belonging to different castes, which carry out separate functions inside the colony. These sharp epigenetic differences present unique opportunities for the experimental dissection of molecular pathways that may be conserved in other organisms, including humans.

Keywords: epigenetics; noncoding RNA; ants; behavior; chromatin

Genetics and epigenetics

In the 19th century, Gregor Mendel discovered that discrete units of biological information, such as those that specify the color and shape of pea seeds, are transmitted from the parents to the progeny according to precise laws of segregation.[1] These units of information, which we now call "genes," are defined segments of the DNA sequence that store the information to synthesize proteins or RNAs that carry out specific functions; for example, one of Mendel's genes encodes a regulator of chlorophyll degradation that causes color changes in the pea seeds.[2] Genes with different DNA sequences (genotypes) encode different proteins that yield different observable traits (phenotypes). After the rediscovery of Mendel's work in the early 20th century, it became apparent that not all inherited traits follow the laws of segregation and that, in some cases, individuals with the same genotype exhibit different phenotypes, which can at times be inherited.[3]

For example, through proper breeding of maize plants, genetically identical individuals can be obtained that differ greatly in pigmentation through the phenomenon of "paramutation."[4] Such inherited traits that cannot be traced to changes in the primary DNA sequence are called "epi-genetic" to indicate that they must rely on inheritance mechanisms beyond Mendel's laws of genetics.[5,6]

Although there are examples of epigenetic inheritance of organism-wide traits, such as agouti coat color in mice[7] and flower geometries in the toadflax *Linaria vulgaris*,[8] the phenomenon is much more pervasive at the cellular level. Epigenetic inheritance of cellular identities is observed in bacteria,[9] yeast,[10] and is absolutely required in multicellular organisms, where the same zygotic genome must epigenetically specify many distinct cell types. This cellular specialization is achieved by the selective activation and repression of cell-specific genes in a highly coordinated spatial and temporal manner. Once the gene expression patterns that determine

doi: 10.1111/j.1749-6632.2011.06363.x

cell identity are set, they are typically inherited by the cell's progeny when it divides, even after the establishing stimulus has subsided.[11] When epigenetic maintenance of cell specialization fails, healthy cells can "lose" their identity and acquire a transformed phenotype that can give rise to cancer.[12]

Research on the molecular mechanisms of epigenetics and their phenotypic counterpart has accelerated exponentially in the last decades. Several molecular pathways have incontrovertibly been linked to epigenetics, and several more have been proposed.[6] It is becoming clear that epigenetic mechanisms do not only control the inheritance of some inconsequential phenotypic trait, but also affect many fundamental and medically relevant biological phenomena, from development to cancer, from regeneration to behavior.

Molecular epigenetics

The epigenetic transmission of phenotypic traits, both molecular from cell to cell and macroscopic from generation to generation, relies on the existence of epigenetic signals: molecular mechanisms that can establish and maintain epigenetic memories. From a conceptual perspective, an epigenetic memory is a form of bistable switch that turns on or off in response to an external stimulus, remains in that state after the stimulus is withdrawn, and is inherited in the same state after cell division. In its simplest form, the switch can consist of one or more feedback loops featuring transcription factors that regulate specific gene transcription and can be transmitted to the cell's progeny. This type of cellular memory is referred to as "trans epigenetics."[6,13] Although this mechanism is widely used in all kingdoms of life, it does not suffice to explain all epigenetic phenomena, especially those in which two identical DNA sequences are treated differently in the same nucleus, as in the case of X chromosome inactivation[14] and parental imprinting.[15] Thus, some epigenetic signals must be encoded in cis; that is to say that they must physically reside on the same molecular structure that encodes the genetic information, the chromosome.

In recent years, epigenetic research has focused on the molecular components of chromatin, the mixture of DNA, RNA, and proteins in which eukaryotic chromosomes are organized inside the nucleus. Epigenetic marks can be deposited directly on the DNA by chemical modification of one of its bases, as in the case of methylation and hydroxymethylation of cytosines,[16,17] or can be engraved in the protein components of chromatin (Fig. 1). In all nucleated cells, DNA is tightly coiled into disc-like structures called nucleosomes that contain two copies each of four basic proteins called histones: H2A, H2B, H3, and H4. The N-terminal and C-terminal tails of the histone proteins protrude from the nucleosome and are extensively modified by enzymes that place or remove chemical marks such as methylation, acetylation, phosphorylation, etc.[18] These marks are recognized by effector proteins that stimulate or repress transcription,[19–21] regulate splicing,[22] participate in DNA replication[23] and repair,[24] or even direct DNA recombination as in the rearrangements of immunoglobulin loci in lymphocytes.[25] Histone marks that are stable enough to be transmitted to the next generation may constitute bona fide epigenetic signals,[26,27] although the extent to which this happens remains controversial. In addition, some histones have alternative isoforms that differ in the primary amino acid sequence and may themselves function as epigenetic signals.[28]

Numerous studies have characterized the distribution of histone marks along the genome in humans and several model organisms, and, in many cases, strong correlations with transcription as well as other chromatin-related processes have been observed.[29] However, we are still far from understanding the meaning and function of many if not all histone marks. This may come partially from a reductionist focus on a handful of histone modifications, although more than 150 have been observed.[30] Bolstered by the recent advances in parallel sequencing technologies and bioinformatics, a systematic annotation of histone marks aimed to determine their genome-wide distribution in humans[31] and model organisms[32] is underway, but only multidisciplinary approaches that combine these genome-wide studies with genetic and biochemical experimentation will ultimately allow us to determine which histone marks carry epigenetic information and which ones are just molecular consequences of transcription and other changes in chromatin.[33]

Noncoding RNAs and chromatin

Regardless of the molecular details of the mechanistic connection between chromatin modifications and transcription, the questions most

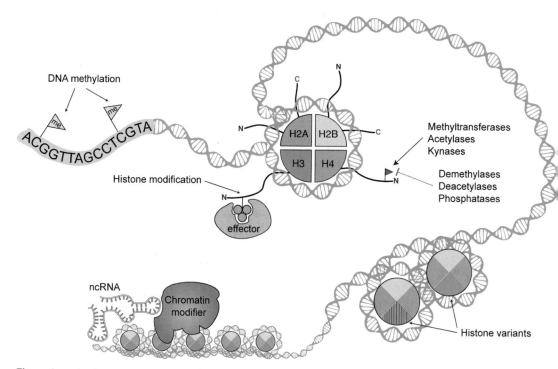

Figure 1. Molecular epigenetics. Schematic depiction of the various molecular factors discussed in the text. DNA methylation and various types of histone modifications are deposited and removed by a large family of chromatin-modifying enzymes with different specificities and often function by recruiting or stabilizing effector proteins that are responsible for downstream functions. Histone variants may also be involved in epigenetic processes. Finally, a popular model envisions ncRNAs as recruitment factors for some chromatin-modifying activities.

relevant for our understanding of epigenetic inheritance concern how they are targeted to specific chromosomal regions and how they are retained after the large-scale disruptions that accompany DNA replication, chromosome condensation, and cell division.

A large family of enzymes and enzymatic complexes known as "histone modifiers" are responsible for placing and removing histone marks,[34] and DNA methyl-transferases modify DNA directly.[35] The identity of many of these enzymes is known, but in very few cases do we fully understand how they are specifically recruited to their chromatin targets. Although direct recruitment by sequence-specific transcription factors is certainly a possibility, it still does not explain how active and repressed chromatin domains are maintained once transcription factor expression wanes, as is often the case during embryonic development,[11] or how two DNA molecules with the same exact sequence can be treated differently by the chromatin-modifying machinery.

One possibility that has attracted much attention in recent years is that noncoding RNAs (ncRNAs) may serve as a regulatory layer between chromatin-modifying complexes and the DNA sequence. Contrary to its original denomination as "junk" DNA,[36] most of the 98% of the human genome devoid of protein-coding potential is transcribed, and a considerable fraction is also conserved in sequence among different species, more than would be expected by chance.[37–39] According to the RNA world hypothesis, the earliest self-replicating systems bearing resemblance to current life forms were entirely based on RNA,[40,41] possibly because it is the most versatile biological polymer. Like proteins, RNA molecules can fold into complex tertiary structures with elaborate surfaces and cavities that can mediate high-specificity interactions and even catalyze biochemical reactions;[42] like DNA, RNA can form Watson–Crick, sequence-specific interactions with other RNAs or with DNA itself. In other words, RNA is fluent in two languages: the elaborate three-dimensional discourse of proteins and the linear genetic code of DNA. Thus, it seems fitting that RNA may act as a molecular bridge—an epigenetic "translator"—between chromatin-modifying proteins and the genome sequence.

Many classes of ncRNAs exist, more than could be accurately described here, and excellent reviews have been published on the topic.[37,43,44] For classification purposes, we distinguish long (>200 nts) ncRNAs, typically transcribed by RNA polymerase II from sequences similar to protein-coding genes,[45] and short (20–40 nts) ncRNAs that originate from different pathways and usually act as guides for protein complexes.[37] Other species with intermediate size may exist, but they are even less characterized.[46,47] Although short ncRNAs contribute to epigenetic memory in yeast[43] and plants,[48,49] very little evidence for similar pathways exists in humans.

A class of long intergenic ncRNAs (lincRNAs) has become the latest focus of epigenetic research. LincRNAs are transcribed by RNA polymerase II and processed similarly to mRNAs for protein-coding genes, but they do not contain open-reading frames. Despite the lack of protein-coding potential, their sequence is more conserved than what would be expected by chance, suggesting a biological function that in most cases remains unknown.[50] So far, lincRNAs have been attributed roles in cancer[51] and pluripotency,[52] possibly via their extensive network of interaction with chromatin modifiers.[53,54] It is estimated that the human genome encodes 5,000–8,000 lincRNAs.[55] On average, they are expressed at lower levels than protein-coding mRNAs and typically in a very tissue-specific fashion. Interestingly, lincRNAs are particularly abundant in the brain, where their expression pattern is often confined to well-defined anatomical regions.[56]

Epigenetic pathways in the brain

All multicellular organisms face the problem of encoding distinct cellular phenotypes in a single genome, and this task appears particularly challenging when we consider brain development and function. The human brain is composed of 10^{11} neurons from hundreds of different types, organized in an outstandingly complex network estimated to have up to 10^{15} connections.[57,58] Neurons are confronted with molecular problems that may seem insurmountable: they must discriminate among a myriad of unwanted connections to establish the correct ones, and they must be capable of retaining impressions of external stimuli in the form of long-term memories, in some cases for decades.[59]

It appears hardly possible that the ~20,000 proteins encoded in the human genome[60] could provide enough complexity to support brain development and function. In some aspects, neuroscientists are faced with a puzzle similar to that faced by immunologists decades ago. The fact that antibodies could be raised against a virtually unlimited number of antigens seemed to challenge Beadle and Tatum's "one gene, one enzyme" hypothesis. How could a limited number of genes encode such a large variety of antibodies? We now know that antibody diversity is generated through V(D)J recombination, a process by which short gene fragments are randomly shuffled to yield billions of combinations.[61] Similarly, at least some degree of combinatorial diversity is involved in brain development and function. For example, *Dscam1* in fruit flies and the clustered protocadherin family in mammals both use alternative splicing to generate unique molecular fingerprints for developing neurons, allowing them to distinguish self from nonself when establishing connections.[62] These neuronal identities and choices of alternative splicing need to persist longer than the turnover rate for RNA, suggesting that epigenetic pathways may be involved. Interestingly, DNA methylation can influence exon selection,[63] offering a potential molecular mechanism for the long-term epigenetic maintenance of alternative splicing choices.

An attractive hypothesis is that the same epigenetic pathways that control transcriptional memory during development may have been co-opted by the animal brain and that they contribute to neuronal identity and diversification, as well as to the long-term stability of neurological memory.[64,65] Differentiated neurons do not divide and cannot transmit information to their progeny; therefore, strictly speaking, long-term transcriptional memory in these cells is not epigenetic. However, Nanney pointed out half a century ago that "the separation of epigenetic regulators into two major classes depending on their stability would separate regulators with fundamentally similar mechanisms into two artificial categories:"[66] that is to say, the very same molecular pathways may control epigenetically inherited traits as well as long-term cellular memory in nondividing cells, such as neurons. Unveiling the relationship between epigenetics and behavior has far-reaching implications for human health, given the known or suspected influence of epigenetic pathways on addiction[67,68] and other psychiatric disorders such as depression and schizophrenia.[69]

Several lines of evidence support the hypothesis that molecular epigenetics has a special role in brain function. Histone acetylation and methylation have both been linked to learning and memory,[70–73] to the point that some small molecule inhibitors of histone deacetylases are under consideration for the treatment of neurodegenerative diseases such as Parkinson's and Alzheimer's.[74] The DNA methylation status of genes involved in brain function can also change rapidly upon neuronal stimulation,[75,76] and the continued expression of DNA methyltransferases in postmitotic forebrain neurons is required for learning and memory in mice.[77] Mutations in *MECP2*, which encodes a protein that binds to methylated DNA, cause Rett's syndrome,[78] a severe form of mental retardation. Similar neurological defects in adult mutant mice can be rescued by expression of wild-type *Mecp2*,[79] suggesting that the pathology does not arise only from developmental defects, but that MECP2 (and likely DNA methylation) is essential for adult brain functions. The monoallelic expression of genes according to their parent of origin is widespread in the mouse brain, with more than 1,300 loci affected.[80] This phenomenon, known as parental imprinting, epitomizes the concept of differential epigenetic regulation: the imprinted (silent) and nonimprinted (active) alleles are usually identical in their DNA sequence and the epigenetic difference is encoded in their methylation status,[15] often accompanied by local changes in chromatin structure and differential transcription of ncRNAs.

As discussed in the previous section, mounting evidence supports a role for ncRNAs in multiple epigenetic pathways, including those affecting chromatin structure (histone modifications) and imprinting. If these epigenetic pathways are central to brain function, it follows that ncRNAs should also play an important role.[81] Indeed, a large fraction of lincRNAs are expressed in the adult brain, and many of them are transcribed from loci also encoding proteins with neuronal functions.[56] Several of the most rapidly evolving human genes are noncoding and exclusively expressed in the brain,[82] and ncRNAs are being implicated in an increasing number of neurological diseases.[83–85] A role for ncRNAs in neuronal plasticity and learning is also supported by studies

in the songbird zebra finch, in which a large number of noncoding transcripts are modulated as the animal listens to another bird's vocalizations.[86] This suggests that the function of ncRNA in the animal brain may not be a recent evolutionary invention and that it may be fruitful to seek evidence of these pathways in organisms less related to humans but more accessible to experimentation.

Epigenetic specification of alternative behaviors in ants

The possibility that epigenetic mechanisms may have been adapted to the specialized functions of the animal brain is highly suggestive but hardly demonstrated. Because of the striking behavioral differences among castes, eusocial insects such as ants, bees, wasps, and termites present unique opportunities to investigate the connection between epigenetic pathways, the brain, and behavior.[87] In fact, studies in the honeybee have pointed at a potential role of DNA methylation in learning and memory.[88]

With more than 20,000 species claiming 15–20% of the terrestrial animal biomass, ants are the most successful eusocial insects and display the largest variety of behavioral repertoires and social structures.[89] Worker ants are often organized in distinct castes that exhibit different morphologies, lifespans, and behaviors.[89] Although a genetic component exists in certain species,[90,91] caste determination seems predominantly guided by environmental stimuli,[92] suggesting that inter-caste differences can be specified epigenetically from the same genetic information, a process that bears a suggestive resemblance to that of cellular differentiation and specialization in multicellular organisms. The genomes of the honeybee and of six ant species are now publicly available,[87,93–97] which should facilitate the progress in our molecular understanding of eusociality and of the epigenetic determination of alternative behaviors.

Two ant species, *Camponotus floridanus* and *Harpegnathos saltator*, exhibit intriguing contrasts in their behavioral flexibility and social organization.[87] *C. floridanus* ants live in large colonies where only the long-lived queen lays fertilized eggs. If the queen is removed or dies, so does the colony. *C. floridanus* workers are divided into minor and major castes, both functionally sterile and differing only in morphology and behavior (Fig. 2A). In contrast, *H. saltator* colonies are small and, although

A

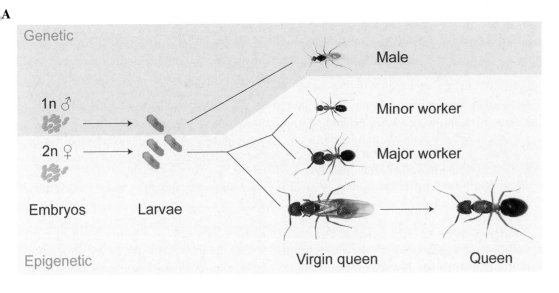

B

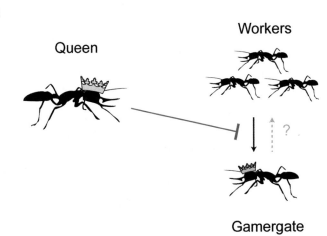

Figure 2. Epigenetic paradigms in ants. (A) In *C. floridanus*, as in all ant species, males develop from haploid embryos (1n, top). All other castes develop from genetically indistinguishable 2n female embryos. In addition to the distinction between reproductives (virgin queen and queen) and nonreproductives, the worker caste in *C. floridanus* is dimorphic, with minor workers mostly responsible for foraging and major workers for defense. (B) In *H. saltator*, workers have the potential to become reproductive gamergates, but this transition (black arrow) is inhibited by the presence of the queen (red line). When the queen dies, a period of intense fighting ensues, after which a few individuals earn reproductive rights. These individuals (gamergates) exhibit dramatic changes in behavior inside the colony. Whether the behavioral changes are reversible (gray dashed arrow) is currently under investigation. Photographs courtesy of Jürgen Liebig.

typically a single queen is present, some workers mate with males from the colony and lay fertilized eggs. If the queen is removed, workers become aggressive toward each other and supremacy fights ensue until a few dominant ants prevail. These individuals undergo a dramatic behavioral and physiological switch to become new functional queens called "gamergates" (Fig. 2B). During this transition, gamergates experience pronounced changes in the expression of genes involved in lifespan regulation, epigenetic pathways, and, possibly, chemical communication.[87] Among the epigenetic regulators, gamergates express higher levels of *SMYD3* and lower levels of *SMYD4* compared to workers. These poorly characterized enzymes are histone modifiers belonging to the methyltransferase family, and their caste-specific expression suggests a role in regulating or implementing the worker–gamergate transition. Compared to workers, gamergates express a large number of small ncRNAs of unknown origin[87] that may also be involved in epigenetic regulation.

Individuals from the two workers castes in *C. floridanus* develop distinct behaviors early during development and maintain them throughout their life, whereas *H. saltator* workers that become gamergates switch on queen-like behaviors as fully developed adults. In both cases, the distinct behaviors arise from indistinguishable genomes and provide compelling experimental settings to investigate the epigenetic regulation of brain function. Genes differentially expressed between *C. floridanus* minor and major workers were enriched for gene ontology annotations related to brain function, including "post-synaptic membrane," "sensory perception of smell," and "neurotransmitter binding,"[87] suggesting that some of the most salient molecular differences between minor and major workers are indeed found in the brain.

The RNA analyzed in this study was obtained from whole organisms, but ant brains can be isolated with relative ease, and this will pave the way to more focused questions about the relationship between genes and behavior. Specifically, the transcriptional profiling of brains from different castes will unveil correlations between the different behavioral outcomes and the underlying molecular phenotypes, such as the expression levels of protein-coding and noncoding RNAs, the genome-wide profiles for histone marks, and changes in DNA methylation. The functional implications of these correlations will need to be investigated with loss- and gain-of-function experiments either through the implementation of genetic tools in ants or by analyzing homologous pathways in more tractable model organisms.

Conclusions and future directions

Among the high-priority goals of modern biology is determining how different cell types are specified from a single genome and what are the molecular underpinnings of higher brain functions such as learning, memory, and behavior. Further progress in our understanding of molecular epigenetics is essential for the former and may help us to achieve the latter.

Although the revolution in parallel sequencing currently allows us to generate comprehensive genome-wide datasets on histone marks, DNA methylation, and protein–RNA interactions, these approaches often emphasize correlation over causation. In other words, although we can accurately assign a certain molecular signature on chromatin to active or repressed regions of the genome, it remains challenging to determine whether it constitutes a true epigenetic mark; that is to say, if it causes the observed state. Yet, this mechanistic insight is essential for our ability to understand and manipulate epigenetic states and will still require tried-and-true biochemical approaches as well as novel technologies.

For example, chemical biology strategies to label protein–protein and protein–RNA interactions in the crowded environment of the nucleus will be needed to untangle the molecular relationships between the various epigenetic factors. More sensitive versions of current sequencing approaches will be required to study the epigenetic signatures of small cell populations or even of individual cells. This will be of critical importance, as exemplified in a recent study that unveiled how an epigenetic switch that is bistable at the single-cell level resulted in a quantitative response at the whole organism level.[98] All new hypotheses will have to be thoroughly tested by loss-of-function and gain-of-function studies. For this, it will be important to artificially introduce targeted changes in a cell's epigenetic state—a form of site-specific epimutagenesis—for example, through the recruitment of enzymatic activities to chromatin via fusion to sequence-specific DNA-binding proteins or by using ncRNAs as guiding devices. Finally, new model organisms, including but not limited to eusocial insects, will provide unexpected new avenues of investigation to dissect the link between epigenetic pathways, brain function, and other high-level organism functions.

Acknowledgments

I thank David Beck, Sergei Koralov, Varun Narendra, and Melissa Phegley for careful reading of this manuscript. I am indebted to Danny Reinberg for constant encouragement, unwavering support, and engaging discussions. My postdoctoral studies have been supported in part by a fellowship from the Helen Hay Whitney Foundation. The ant epigenetic project is funded by the Howard Hughes Medical

Institute Collaborative Innovation Award #2009005 to Shelley Berger, Jürgen Liebig, and Danny Reinberg.

Conflicts of interest

The author declares no conflicts of interest.

References

1. Mendel, G. 1866. Versuche über Pflanzen-Hybriden. *Verh. Naturforsch. Ver. Brünn* **4:** 3–47.

2. Sato, Y., R. Morita, M. Nishimura, H. Yamaguchi & M. Kusaba. 2007. Mendel's green cotyledon gene encodes a positive regulator of the chlorophyll-degrading pathway. *Proc. Natl. Acad. Sci. USA* **104:** 14169–14174.

3. Holliday, R. 2006. Epigenetics: a historical overview. *Epigenetics* **1:** 76–80.

4. Chandler, V.L. 2007. Paramutation: from maize to mice. *Cell* **128:** 641–645.

5. Nise, M.S., P. Falaturi & T.C. Erren. 2010. Epigenetics: origins and implications for cancer epidemiology. *Med. Hypotheses.* **74:** 377–382.

6. Bonasio, R., S. Tu & D. Reinberg. 2010. Molecular signals of epigenetic states. *Science* **330:** 612–616.

7. Morgan, H.D., H.G. Sutherland, D.I. Martin & E. Whitelaw. 1999. Epigenetic inheritance at the agouti locus in the mouse. *Nat. Genet.* **23:** 314–318.

8. Cubas, P., C. Vincent & E. Coen. 1999. An epigenetic mutation responsible for natural variation in floral symmetry. *Nature* **401:** 157–161.

9. Casadesus, J. & D. Low. 2006. Epigenetic gene regulation in the bacterial world. *Microbiol. Mol. Biol. Rev.* **70:** 830–856.

10. Grewal, S.I. & A.J. Klar. 1996. Chromosomal inheritance of epigenetic states in fission yeast during mitosis and meiosis. *Cell* **86:** 95–101.

11. Ringrose, L. & R. Paro. 2004. Epigenetic regulation of cellular memory by the Polycomb and Trithorax group proteins. *Annu. Rev. Genet.* **38:** 413–443.

12. Berdasco, M. & M. Esteller. 2010. Aberrant epigenetic landscape in cancer: how cellular identity goes awry. *Dev. Cell* **19:** 698–711.

13. Moazed, D. 2011. Mechanisms for the inheritance of chromatin states. *Cell* **146:** 510–518.

14. Payer, B. & J.T. Lee. 2008. X chromosome dosage compensation: how mammals keep the balance. *Annu. Rev. Genet.* **42:** 733–772.

15. Ferguson-Smith, A.C. 2011. Genomic imprinting: the emergence of an epigenetic paradigm. *Nat. Rev. Genet.* **12:** 565–575.

16. Bird, A. 2002. DNA methylation patterns and epigenetic memory. *Genes. Dev.* **16:** 6–21.

17. Ndlovu, M.N., H. Denis & F. Fuks. 2011. Exposing the DNA methylome iceberg. *Trends. Biochem. Sci.* **36:** 381–387.

18. Campos, E. & D. Reinberg. 2009. Histones: annotating Chromatin. *Annu. Rev. Genet.* **43:** 559–599.

19. Taverna, S.D. *et al.* 2006. Yng1 PHD finger binding to H3 trimethylated at K4 promotes NuA3 HAT activity at K14 of H3 and transcription at a subset of targeted ORFs. *Mol. Cell* **24:** 785–796.

20. Vermeulen, M. *et al.* 2007. Selective anchoring of TFIID to nucleosomes by trimethylation of histone H3 lysine 4. *Cell* **131:** 58–69.

21. Simon, J.A. & R.E. Kingston. 2009. Mechanisms of polycomb gene silencing: knowns and unknowns. *Nat. Rev. Mol. Cell Biol.* **10:** 697–708.

22. Sims, R.J. *et al.* 2005. Human but not yeast CHD1 binds directly and selectively to histone H3 methylated at lysine 4 via its tandem chromodomains. *J. Biol. Chem.* **280:** 41789–41792.

23. Chakraborty, A., Z. Shen & S.G. Prasanth. 2011. "ORCanization" on heterochromatin: linking DNA replication initiation to chromatin organization. *Epigenetics* **6:** 665–670.

24. Oda, H. *et al.* 2010. Regulation of the histone H4 monomethylase PR-Set7 by CRL4(Cdt2)-mediated PCNA-dependent degradation during DNA damage. *Mol. Cell* **40:** 364–376.

25. Matthews, A.G. *et al.* 2007. RAG2 PHD finger couples histone H3 lysine 4 trimethylation with V(D)J recombination. *Nature* **450:** 1106–1110.

26. Margueron, R. *et al.* 2009. Role of the polycomb protein EED in the propagation of repressive histone marks. *Nature* **461:** 762–767.

27. Hansen, K.H. *et al.* 2008. A model for transmission of the H3K27me3 epigenetic mark. *Nat. Cell Biol.* **10:** 1291–1300.

28. Talbert, P.B. & S. Henikoff. 2010. Histone variants—ancient wrap artists of the epigenome. *Nat. Rev. Mol. Cell Biol.* **11:** 264–275.

29. Guenther, M.G., S.S. Levine, L.A. Boyer, *et al.* 2007. A chromatin landmark and transcription initiation at most promoters in human cells. *Cell* **130:** 77–88.

30. Tan, M. *et al.* 2011. Identification of 67 histone marks and histone lysine crotonylation as a new type of histone modification. *Cell* **146:** 1016–1028.

31. Ernst, J. *et al.* 2011. Mapping and analysis of chromatin state dynamics in nine human cell types. *Nature* **473:** 43–49.

32. Roy, S. *et al.* 2010. Identification of functional elements and regulatory circuits by Drosophila modENCODE. *Science* **330:** 1787–1797.

33. Limb, J. & B.M. Turner. 2007. Defining an epigenetic code. *Nat. Cell Biol.* **9:** 2–6.

34. Gardner, K.E., C.D. Allis & B.D. Strahl. 2011. Operating on chromatin, a colorful language where context matters. *J. Mol. Biol.* **409:** 36–46.

35. Goll, M.G. & T.H. Bestor. 2005. Eukaryotic cytosine methyltransferases. *Annu. Rev. Biochem.* **74:** 481–514.

36. Makalowski, W. 2003. Not junk after all. *Science* **300:** 1246–1247.

37. Jacquier, A. 2009. The complex eukaryotic transcriptome: unexpected pervasive transcription and novel small RNAs. *Nat. Rev. Genet.* **10:** 833–844.

38. Carninci, P., J. Yasuda & Y. Hayashizaki. 2008. Multifaceted mammalian transcriptome. *Curr. Opin. Cell Biol.* **20:** 274–280.

39. Carninci, P. *et al.* 2005. The transcriptional landscape of the mammalian genome. *Science* **309:** 1559–1563.

40. Crick, F.H. 1968. The origin of the genetic code. *J. Mol. Biol.* **38:** 367–379.

41. Orgel, L.E. 1968. Evolution of the genetic apparatus. *J. Mol. Biol.* **38:** 381–393.

42. Lilley, D.M. 2011. Mechanisms of RNA catalysis. *Philos. Trans. R. Soc. Lond. B. Biol. Sci.* **366:** 2910–2917.

43. Moazed, D. 2009. Small RNAs in transcriptional gene silencing and genome defence. *Nature* **457:** 413–420.

44. Pauli, A., J.L. Rinn & A.F. Schier. 2011. Non-coding RNAs as regulators of embryogenesis. *Nat. Rev. Genet.* **12:** 136–149.

45. Mercer, T., M. Dinger & J. Mattick. 2009. Long non-coding RNAs: insights into functions. *Nat. Rev. Genet.* **10:** 155–159.

46. Kanhere, A. *et al.* 2010. Short RNAs are transcribed from repressed polycomb target genes and interact with polycomb repressive complex-2. *Mol. Cell* **38:** 675–688.

47. Kapranov, P. *et al.* 2007. RNA maps reveal new RNA classes and a possible function for pervasive transcription. *Science* **316:** 1484–1488.

48. Matzke, M., T. Kanno, B. Huettel, L. Daxinger & A.J. Matzke. 2007. Targets of RNA-directed DNA methylation. *Curr. Opin. Plant Biol.* **10:** 512–519.

49. Autran, D. *et al.* 2011. Maternal epigenetic pathways control parental contributions to Arabidopsis early embryogenesis. *Cell* **145:** 707–719.

50. Guttman, M. *et al.* 2009. Chromatin signature reveals over a thousand highly conserved large non-coding RNAs in mammals. *Nature* **458:** 223–227.

51. Gupta, R.A. *et al.* 2010. Long non-coding RNA HOTAIR reprograms chromatin state to promote cancer metastasis. *Nature* **464:** 1071–1076.

52. Loewer, S. *et al.* 2010. Large intergenic non-coding RNA-RoR modulates reprogramming of human induced pluripotent stem cells. *Nat. Genet.* **42:** 1113–1117.

53. Koziol, M.J. & J.L. Rinn. 2010. RNA traffic control of chromatin complexes. *Curr. Opin. Genet. Dev.* **20:** 142–148.

54. Khalil, A. *et al.* 2009. Many human large intergenic noncoding RNAs associate with chromatin-modifying complexes and affect gene expression. *Proc. Natl. Acad. Sci. USA* **106:** 11667–11672.

55. Cabili, M.N. *et al.* 2011. Integrative annotation of human large intergenic noncoding RNAs reveals global properties and specific subclasses. *Genes Dev.* **25:** 1915–1927.

56. Mercer, T.R., M.E. Dinger, S.M. Sunkin, *et al.* 2008. Specific expression of long noncoding RNAs in the mouse brain. *Proc. Natl. Acad. Sci. USA* **105:** 716–721.

57. Azevedo, F.A. *et al.* 2009. Equal numbers of neuronal and nonneuronal cells make the human brain an isometrically scaled-up primate brain. *J. Comp. Neurol.* **513:** 532–541.

58. Williams, R.W. & K. Herrup. 1988. The control of neuron number. *Annu. Rev. Neurosci.* **11:** 423–453.

59. Squire, L.R. 1987. *Memory and Brain.* Oxford University Press. New York.

60. Consortium, I.H.G.S. 2004. Finishing the euchromatic sequence of the human genome. *Nature* **431:** 931–945.

61. Jung, D., C. Giallourakis, R. Mostoslavsky, *et al.* 2006. Mechanism and control of V(D)J recombination at the immunoglobulin heavy chain locus. *Annu. Rev. Immunol.* **24:** 541–570.

62. Zipursky, S.L. & J.R. Sanes. 2010. Chemoaffinity revisited: dscams, protocadherins, and neural circuit assembly. *Cell* **143:** 343–353.

63. Shukla, S. *et al.* 2011. CTCF-promoted RNA polymerase II pausing links DNA methylation to splicing. *Nature* **479:** 74–79.

64. Day, J.J. & J.D. Sweatt. 2011. Epigenetic mechanisms in cognition. *Neuron* **70:** 813–829.

65. Dulac, C. 2010. Brain function and chromatin plasticity. *Nature* **465:** 728–735.

66. Nanney, D.L. 1958. Epigenetic control systems. *Proc. Natl. Acad. Sci. USA* **44:** 712–717.

67. Robison, A.J. & E.J. Nestler. 2011. Transcriptional and epigenetic mechanisms of addiction. *Nat. Rev. Neurosci.* **12:** 623–637.

68. Maze, I. *et al.* 2010. Essential role of the histone methyltransferase G9a in cocaine-induced plasticity. *Science* **327:** 213–216.

69. Tsankova, N., W. Renthal, A. Kumar & E.J. Nestler. 2007. Epigenetic regulation in psychiatric disorders. *Nat. Rev. Neurosci.* **8:** 355–367.

70. Peleg, S. *et al.* 2010. Altered histone acetylation is associated with age-dependent memory impairment in mice. *Science* **328:** 753–756.

71. Fischer, A., F. Sananbenesi, X. Wang *et al.* 2007. Recovery of learning and memory is associated with chromatin remodelling. *Nature* **447:** 178–182.

72. Guan, J.S. *et al.* 2009. HDAC2 negatively regulates memory formation and synaptic plasticity. *Nature* **459:** 55–60.

73. Gupta, S. *et al.* 2010. Histone methylation regulates memory formation. *J. Neurosci.* **30:** 3589–3599.

74. Hahnen, E. *et al.* 2008. Histone deacetylase inhibitors: possible implications for neurodegenerative disorders. *Expert. Opin. Investig. Drugs* **17:** 169–184.

75. Guo, J.U. *et al.* 2011. Neuronal activity modifies the DNA methylation landscape in the adult brain. *Nat. Neurosci.* **14:** 1345–1351.

76. Martinowich, K. *et al.* 2003. DNA methylation-related chromatin remodeling in activity-dependent BDNF gene regulation. *Science* **302:** 890–893.

77. Feng, J. *et al.* 2010. Dnmt1 and Dnmt3a maintain DNA methylation and regulate synaptic function in adult forebrain neurons. *Nat. Neurosci.* **13:** 423–430.

78. Amir, R.E. *et al.* 1999. Rett syndrome is caused by mutations in X-linked MECP2, encoding methyl-CpG-binding protein 2. *Nat. Genet.* **23:** 185–188.

79. Luikenhuis, S., E. Giacometti, C.F. Beard & R. Jaenisch. 2004. Expression of MeCP2 in postmitotic neurons rescues Rett syndrome in mice. *Proc. Natl. Acad. Sci. USA* **101:** 6033–6038.

80. Gregg, C. *et al.* 2010. High-resolution analysis of parent-of-origin allelic expression in the mouse brain. *Science* **329:** 643–648.

81. Mehler, M.F. & J.S. Mattick. 2007. Noncoding RNAs and RNA editing in brain development, functional diversification, and neurological disease. *Physiol. Rev.* **87:** 799–823.

82. Pollard, K.S. *et al.* 2006. An RNA gene expressed during cortical development evolved rapidly in humans. *Nature* **443:** 167–172.

83. Sopher, B.L. *et al.* 2011. CTCF regulates ataxin-7 expression through promotion of a convergently transcribed, antisense noncoding RNA. *Neuron* **70:** 1071–1084.

84. St Laurent, G., 3rd, M.A. Faghihi & C. Wahlestedt. 2009. Non-coding RNA transcripts: sensors of neuronal stress, modulators of synaptic plasticity, and agents of change in the onset of Alzheimer's disease. *Neurosci. Lett.* **466:** 81–88.

85. Faghihi, M.A. *et al.* 2008. Expression of a noncoding RNA is elevated in Alzheimer's disease and drives rapid feed-forward regulation of beta-secretase. *Nat. Med.* **14:** 723–730.

86. Warren, W.C. *et al.* 2010. The genome of a songbird. *Nature* **464:** 757–762.

87. Bonasio, R. *et al.* 2010. Genomic comparison of the ants Camponotus floridanus and Harpegnathos saltator. *Science* **329:** 1068–1071.

88. Lockett, G.A., P. Helliwell & R. Maleszka. 2010. Involvement of DNA methylation in memory processing in the honey bee. *Neuroreport* **21:** 812–816.

89. Hölldobler, B. & E.O. Wilson. 1990. *The Ants.* Harvard University Press. Cambridge, MA.

90. Sirvio, A., P. Pamilo, R.A. Johnson, *et al.* 2011. Origin and evolution of the dependent lineages in the genetic caste determination system of Pogonomyrmex ants. *Evolution* **65:** 869–884.

91. Frohschammer, S. & J. Heinze. 2009. A heritable component in sex ratio and caste determination in a Cardiocondyla ant. *Front. Zool.* **6:** 27.

92. Wheeler, D.E. 1986. Developmental and physiological determinants of caste in social hymenoptera: evolutionary implications. *Am. Naturalist* **128:** 13–34.

93. Smith, C.D. *et al.* 2011. Draft genome of the globally widespread and invasive Argentine ant (Linepithema humile). *Proc. Natl. Acad. Sci. USA* **108:** 5673–5678.

94. Smith, C.R. *et al.* 2011. Draft genome of the red harvester ant Pogonomyrmex barbatus. *Proc. Natl. Acad. Sci. USA* **108:** 5667–5672.

95. Wurm, Y. *et al.* 2011. The genome of the fire ant Solenopsis invicta. *Proc. Natl. Acad. Sci. USA* **108:** 5679–5684.

96. Suen, G. *et al.* 2011. The genome sequence of the leaf-cutter ant Atta cephalotes reveals insights into its obligate symbiotic lifestyle. *PLoS. Genet.* **7:** e1002007.

97. Consortium, H.G.S. 2006. Insights into social insects from the genome of the honeybee Apis mellifera. *Nature* **443:** 931–949.

98. Angel, A., J. Song, C. Dean & M. Howard. 2011. A Polycomb-based switch underlying quantitative epigenetic memory. *Nature* **476:** 105–108.

Ann. N.Y. Acad. Sci. ISSN 0077-8923

ANNALS OF THE NEW YORK ACADEMY OF SCIENCES

Issue: *Blavatnik Awards for Young Scientists*

Lighting up the brain's reward circuitry

Mary Kay Lobo

Department of Anatomy and Neurobiology, University of Maryland School of Medicine, Baltimore, Maryland

Address for correspondence: Mary Kay Lobo, Department of Anatomy and Neurobiology, University of Maryland School of Medicine, 20 Penn Street, Baltimore, MD 21201. mklobo@umaryland.edu

The brain's reward circuit is critical for mediating natural reward behaviors including food, sex, and social interaction. Drugs of abuse take over this circuit and produce persistent molecular and cellular alterations in the brain regions and their neural circuitry that make up the reward pathway. Recent use of optogenetic technologies has provided novel insights into the functional and molecular role of the circuitry and cell subtypes within these circuits that constitute this pathway. This perspective will address the current and future use of light-activated proteins, including those involved in modulating neuronal activity, cellular signaling, and molecular properties in the neural circuitry mediating rewarding stimuli and maladaptive responses to drugs of abuse.

Keywords: light-activated proteins; circuitry; neuron; optogenetic

Introduction

The reward pathway is crucial for an animal's interaction with its environment to promote natural rewards including feeding, sex, and social interaction. Drugs of abuse commandeer this system and exert powerfully rewarding properties through long-lasting molecular and cellular changes on the reward pathway.[1] In addition, other neuropsychiatric disorders, including depression and schizophrenia, result in pathological changes in this circuit.[2,3] Until this past decade, insight into the functional role of the neural circuitry and the specific cell types in these circuits (Fig. 1) in neuropsychiatric disorders has been constrained based on the tools available to neuroscientists. This past decade has seen a tremendous explosion in technologies available to define and manipulate neural circuits and specific cell subtypes within those circuits at the level of activity and molecular signaling on a spatiotemporal scale.

First, neuroscientists can define and identify specific cell types using fluorescent reporter mice, including bacterial artificial chromosome (BAC) transgenic mice and Brainbow mice.[4,5] Second, neuroscientists can either overexpress or knockout specific molecules using cell-type specific transgenic or viral methodologies.[6,7] Third, neuroscientists are

exploiting novel optogenetic tools that consist of expressing light-activated proteins in the central nervous system and then regulating these proteins on a precise spatiotemporal scale with light *in vivo.*[7,8]

The ability to manipulate functions in cell types is incredibly valuable when studying reward pathways, given the cellular heterogeneity and the dynamic properties exhibited during reward-related neurobiology. For example, the nucleus accumbens (NAc), a major reward nucleus, comprises multiple neuronal cell types and receives inputs from multiple brain reward regions, including dopamine inputs from the ventral tegmental area (VTA) and glutamatergic inputs from limbic cortical regions, and the amygdala, hypothalamus, and hippocampus.[9]

The other hallmark of light-activated proteins is the temporal resolution with which they can be "switched on" and "switched off" by exposure to light. The neurobiology of reward behaviors is dynamic, and thus manipulating neuronal activity, neuronal signaling, or molecular properties at specific time points is incredibly advantageous. For example, acute exposure to psychostimulants causes robust induction of c-Fos (a genetic marker of neuronal firing) in the NAc, whereas chronic psychostimulant exposure desensitizes c-Fos induction. However, context-dependent re-exposure to

doi: 10.1111/j.1749-6632.2011.06368.x

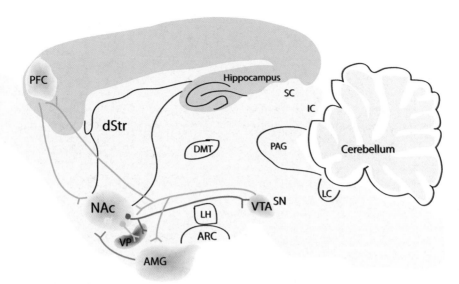

Figure 1. Illustration of the circuitry and the specific cell types within this circuit that make up the brain's reward pathway. The NAc (nucleus accumbens) is an important site for integration, as it receives multiple glutamatergic inputs from the prefrontal cortex (PFC), amygdala (AMG), ventral tegmental area (VTA), and other brain regions not shown, as well as dopaminergic input from the VTA. The NAc is composed of multiple cell subtypes, including D_1^+ MSNs (red), which send collaterals to the ventral pallidum (VP) and project to midbrain regions including the VTA; D_2^+ MSNs (blue), which send collaterals to nearby pallidum regions including VP; and ChAT interneurons (yellow), a subset of local interneurons in the NAc.

psychostimulants induces a subset of c-Fos–activated NAc neurons, and these c-Fos neurons are crucial for mediating context-dependent behavioral sensitization.[10,11] Optogenetic technologies make possible the ability to manipulate protein activity during specific time points, thus allowing precise spatio-temporal regulation of cell types involved in behavioral responses, for example, in the study of addiction. For example, one can silence the c-Fos–activated neurons during psychostimulant context-dependent sensitization (after chronic exposure to drugs of abuse) or one can activate these neurons without the psychostimulant to mimic drug-induced neurobiological responses to address the role of these neurons in behavioral outputs. The ability to precisely control neuronal activity on a spatiotemporal scale to block or mimic a drug stimulus at crucial time points can help to better define the actions of drugs of abuse on particular cell types, for example, during progression from the acute rewarding behavioral response to the chronic addicted behavioral response. In addition, although current overexpression systems have provided important insights into the molecular mechanisms underlying addiction, they lack the precise temporal resolution afforded by optogenetics. Current and future studies using optically activated signaling molecules or transcription factors can better elucidate a given molecule's role at precise time points in chronic disease processes.

Optogenetics

Optogenetics, as the name implies, involves a combination of light optics and genetic expression of light-sensitive proteins that can be targeted to living mammalian cell types and controlled with light on a spatiotemporal scale. This technology was pioneered by Gero Miesenböck, whose group expressed a combination of multiple *Drosophila* photoreceptor genes, which they term *chARGe* to activate hippocampal primary neuron cultures with light.[12] Other pioneers in the field, Ehud Isacoff, Richard Kramer, and Dirk Tauner, developed UV light–isomerizable chemicals linked to a genetically modified channel.[13,14] Finally, Karl Deisseroth, who coined the term *optogenetics*, and Ed Boyden were the first to deliver naturally occurring microbial opsins to mammalian nerve cells.[15] They originally expressed channelrhodopsin-2 (ChR2, a blue light activated cation channel derived from green algae *Chlamydomonas reinhardtii*, Fig. 2) to induce neuronal excitation in mammalian neurons[15] and

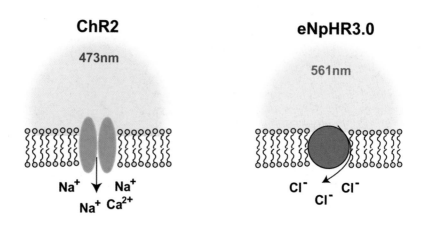

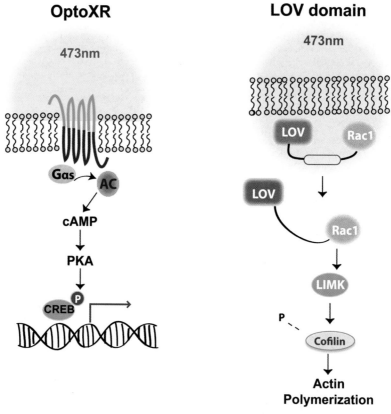

Figure 2. Illustration of four types of light-activated molecules. Channelrhodopsin-2 (ChR2), the blue (473 nm) light activated cation channel, is used to activate neuronal firing. The newest version of halorhodopsin, eNpHR3.0, a yellow (561 nm) light activated chloride pump is used to silence neurons. OptoXRs are G protein-coupled receptor (GPCR)/rhodopsin chimeras that when activated by blue light promote GPCR signaling, eventually leading to activation of transcription factors such as cAMP response element-binding (CREB), which can regulate gene expression in the nucleus. Finally, the LOV domain when activated by blue light can release steric hindrance of a molecule to which it is tethered: in this example, Rac1, allowing the molecule to activate downstream signaling molecules, which eventually alters cellular functions.

then controlled behavior in *Caenorhabditis elegans* by altering neuronal activity with ChR2 and NpHR (halorhodopsin, a yellow light activated chloride pump from a halobacteria *Natronomonas pharaonis*, Fig. 2), which has been effective for yellow light hyperpolarization to silence neurons.[16,17] Shortly thereafter, they and other groups used these microbial opsins to control neuronal activity in rodents.[18–20] The advantage of these microbial opsins is that they allow for a single-component approach involving a light-activated protein within a single gene. With the endogenous presence of an essential organic cofactor retinal in all vertebrate tissues, these microbial opsins can be used without the need for additional exogenous chemicals.[7,8,18] In particular, Karl Deisseroth's group provided tremendous advancement to the neuroscience community by developing multiple light-activated genetic constructs for use in freely moving mammals, which can be easily achieved due to the single-component nature of these tools, and has made this technology widely accessible to the neuroscience community.[7,8,18]

Channelrhodopsins and halorhodopsins

The most extensively used light-activated proteins are channelrhodopsins, which are light-gated cation channels. The most commonly used protein is ChR2 (Fig. 2) in particular ChR2(H134R), a ChR2 variant with a mutation at histidine 134 to arginine to improve steady-state current size.[18] Indeed, ChR2, has been used in multiple behavioral paradigms providing new insights into disorders of the brain including addiction, anxiety, depression, and Parkinson's disease.[21–25] However, other channelrhodopsins, including blue light activated cation channel ChR1 (channelrhodopsin-1) and red light activated VChR1 cation channel (channelrhodopsin-1 from *Volvox carteri*),[26] have been identified, and these three original channelrhodopsins have subsequently been genetically modified for improved kinetics.[7,8] Thus, in the near future it is likely there will be more publications using these newer variants to help fully understand the neurobiology of psychiatric disorders.

Many groups are silencing neurons using microbial opsins, such as halorodopsins (noted above), and light-activated proton pumps, including archeorhodopsins and Mac.[27–29] Deisseroth and colleagues are working with the third generation of NpHR, eNpHR3.0 (Fig. 2), which possesses larger photocurrents, improved expression in cell membranes and processes, and the ability to be activated in the far red light.[28] In addition, Boyden and colleagues are making use of ArchT (an archeorhodopsin), and have recently used this opsin to silence cortical neurons in nonhuman primates.[29]

Photoactivatable G protein–coupled receptors and intracellular molecules

For a molecular biologist studying molecular mechanisms of neuropsychiatric disorders, the ability to switch on and off molecular players within a defined time frame is extremely promising given the dynamic role many molecules play in these diseases. Fortunately, many groups are shifting beyond solely modulating neuronal activity with light-sensitive proteins to developing and using light-sensitive signaling molecules. The Deisseroth group developed G protein-coupled receptor (GPCR) opsin chimeras, termed *OptoXRs* (Fig. 2), in which the intracellular loops of rhodopsin are replaced with a specific GPCR.[30] More recently, Oh *et al.* produced a chimera of rhodopsin and the serotonergic GPCR $G_{i/o}$-coupled 5-HT$_{1A}$,[31] which will be extremely useful, given the role of serotonergic signaling in neuropsychiatric diseases, including depression and anxiety.

Finally, other groups are using naturally occurring light-activated intracellular signaling molecules or are engineering such molecules. Currently, photoactivated nucleotidyl cyclases from bacteria are being used to manipulate cAMP and cGMP levels in cells.[32,33] Klaus Hahn's group is exploiting the light-oxygen-voltage (LOV) protein domain from *Avena sativa* phototropin, which allosterically inhibits proteins fused to this domain in the dark, and upon blue-light exposure a helix linking LOV to each protein unwinds and thus allows the protein to interact with its binding proteins.[34–36] Rac1 plays an important role in regulating actin cytoskeletal dynamics and Hahn's group showed that photoactivatable-Rac1 (PA-Rac1, Fig. 2) in the presence of blue light can alter cell structural dynamics and redirect migrating cells in drosophila.[35,36] Our group is using PA-Rac1 *in vivo* in the rodent brain to investigate the role of Rac1 in cocaine behaviors and cocaine-induced cytoskeleton remodeling.[37] Finally, another group has used the LOV domain to target transcriptional regulation by fusing LOV to the *Escherichia coli* trp repressor (termed *LovTAP*), and this fusion protein selectively binds DNA when illuminated with blue light.[38] The ability to translate this to mammals and create photo-switchable molecules that temporally regulate transcriptional activity is a very exciting prospect for neuroscientists who are trying to elucidate transcriptional alterations in neuropsychiatric diseases.

In recent years, optogenetic tools have been valuable to the drug abuse field, leading to novel understanding of the role of specific cell types and circuitry in the neurobiology of reward. This review will address the current state of optogenetics in reward circuitry but also delve into future paths of focus, particularly those aimed at targeting molecular machinery.

The nucleus accumbens—a site for integration

The NAc is a key integrator of rewarding stimuli and maladaptive responses to drugs of abuse given the multiple inputs converging upon the NAc, which include glutamatergic inputs from the amygdala (emotional content), hippocampus (contextual information), and prefrontal cortex (executive/cognitive information), as well as dopaminergic inputs from the VTA, motivational content (Fig. 1). The NAc is further diversified by the multiple cell subtypes and efferents to other brain regions. The NAc is composed mainly ($\sim$95%) of medium spiny neurons (MSNs), with two main subtypes differentiated by their enrichment of many genes[39,40] and their projections through subcortical structures (Fig. 1).[41] The two MSNs are most notably enriched in dopamine receptor 1 (D_1) or dopamine receptor 2 (D_2) and these D_1^+ and D_2^+ MSNs[40] in both the NAc (ventral striatum) and dorsal striatum play balanced and antagonistic roles in motor and reward-related behaviors.[22,25,42]

Controlling GPCR signaling and activity in NAc neurons

Given the key role of the NAc in rewarding stimuli and maladaptive responses to drugs of abuse, the current optogenetic efforts investigating rewarding behaviors is focused on the NAc or regions projecting to the NAc. The first optogenetic study of rewarding phenomena used OptoXRs (Figs. 2 and 3) to investigate the effect of manipulating biochemical signaling pathways in NAc via a conditioned place preference (CPP) paradigm.[30]

In this paradigm rodents are first conditioned with a stimulus (i.e., optogenetic stimulus, drug injection) in one context-specific chamber and then conditioned to a second context-specific chamber for control purposes (i.e., no optogenetic stimulus, saline injection). Subsequently, rodents are tested for time spent in each chamber, and any additional time spent in the stimulus chamber is an indirect measure of reward. Airan and colleagues expressed the G_q-coupled α_{1a}-adrenergic receptor (opto-α_1AR) and the G_s-coupled β_2-adrenergic receptor (opto-β_2 AR) optoXRs in NAc and demonstrated enhanced place preference for the blue light-paired chamber after activation of opto-α_1AR but not opto-β_2 AR or ChR2.[30] This work demonstrated a unique role for the α_{1a}-adrenergic receptor in mediating rewarding phenomena, which substantiates evaluation of pharmacotherapeutics aimed at this GPCR in treating drug abuse. Furthermore, this study demonstrated a requirement for intracellular signaling in the NAc to achieve rewarding responses, because neuronal activation with ChR2 alone did not induce place preference. Future studies using other OptoXRs for GPCRs differentially expressed in the NAc neurons, in particular the projection neurons, which are well known for their differential expression of GPCRs including dopamine receptors, would be advantageous for understanding cell type–specific GPCR signaling in addiction behaviors.

Recently, we investigated the role of cell type specificity in NAc using optogenetic control by specifically pairing ChR2 activation with cocaine CPP. In particular, we examined the two MSN subtypes using Cre reporters lines (D_1-Cre or D_2-Cre) combined with a conditional ChR2 AAV virus double-inverted open reading frame (DIO)-ChR2-EYFP, in which ChR2 is only expressed in the presence of Cre-recombinase, and used herpes simplex virus (HSV)-ChR2-mCherry to activate all neurons in the NAc (Fig. 3). We showed that optogenetic control of total NAc and D_1^+ MSN neuronal firing potentiates place preference for cocaine, whereas optogenetic control of D_2^+ MSN firing attenuated cocaine place preference.[25] Similar to the Airan finding that ChR2 without pairing to a drug does not affect place preference, we also found that selective activation of D_1^+ or D_2^+ MSNs without cocaine pairing has no effect on CPP. These two studies illustrate the necessity of biochemical signaling (i.e., the direct activation of a GPCR or cocaine's

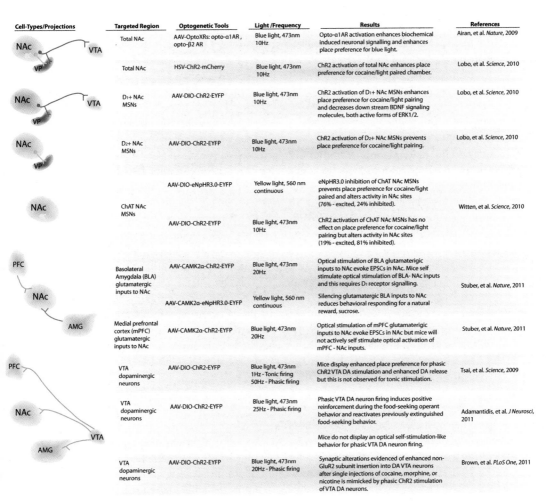

Cell-Types/Projections	Targeted Region	Optogenetic Tools	Light /Frequency	Results	References
	Total NAc	AAV-OptoXRs: opto-α1AR, opto-β2 AR	Blue light, 473nm 10Hz	Opto-α1AR activation enhances biochemical induced neuronal signalling and enhances place preference for blue light.	Airan, et al. *Nature*, 2009
	Total NAc	HSV-ChR2-mCherry	Blue light, 473nm 10Hz	ChR2 activation of total NAc enhances place preference for cocaine/light paired chamber.	Lobo, et al. *Science*, 2010
	D1+ NAc MSNs	AAV-DIO-ChR2-EYFP	Blue light, 473nm 10Hz	ChR2 activation of D1+ NAc MSNs enhances place preference for cocaine/light pairing and decreases down stream BDNF signaling molecules, both active forms of ERK1/2.	Lobo, et al. *Science*, 2010
	D2+ NAc MSNs	AAV-DIO-ChR2-EYFP	Blue light, 473nm 10Hz	ChR2 activation of D2+ NAc MSNs prevents place preference for cocaine/light pairing.	Lobo, et al. *Science*, 2010
	ChAT NAc MSNs	AAV-DIO-eNpHR3.0-EYFP	Yellow light, 560 nm continuous	eNpHR3.0 inhibition of ChAT NAc MSNs prevents place preference for cocaine/light paired and alters activity in NAc sites (76% - excited, 24% inhibited).	Witten, et al. *Science*, 2010
		AAV-DIO-ChR2-EYFP	Blue light, 473nm 10Hz	ChR2 activation of ChAT NAc MSNs has no effect on place preference for cocaine/light pairing but alters activity in NAc sites (19% - excited, 81% inhibited).	
	Basolateral Amygdala (BLA) glutamatergic inputs to NAc	AAV-CAMK2α-ChR2-EYFP	Blue light, 473nm 20Hz	Optical stimulation of BLA glutamatergic inputs to NAc evoke EPSCs in NAc. Mice self stimulate optical stimulation of BLA- NAc inputs and this requires D1 receptor signalling.	Stuber, et al. *Nature*, 2011
		AAV-CAMK2α-eNpHR3.0-EYFP	Yellow light, 560 nm continuous	Silencing glutamatergic BLA inputs to NAc reduces behavioral responding for a natural reward, sucrose.	
	Medial prefrontal cortex (mPFC) glutamatergic inputs to NAc	AAV-CAMK2α-ChR2-EYFP	Blue light, 473nm 20Hz	Optical stimulation of mPFC glutamatergic inputs to NAc evoke EPSCs in NAc but mice will not actively self stimulate optical activation of mPFC - NAc inputs.	Stuber, et al. *Nature*, 2011
	VTA dopaminergic neurons	AAV-DIO-ChR2-EYFP	Blue light, 473nm 1Hz - Tonic firing 50Hz - Phasic firing	Mice display enhanced place preference for phasic ChR2 VTA DA stimulation and enhanced DA release but this is not observed for tonic stimulation.	Tsai, et al. *Science*, 2009
	VTA dopaminergic neurons	AAV-DIO-ChR2-EYFP	Blue light, 473nm 25Hz - Phasic firing	Phasic VTA DA neuron firing induces positive reinforcement during the food-seeking operant behavior and reactivates previously extinguished food-seeking behavior.	Adamantidis, et al. *J Neurosci*, 2011
				Mice do not display an optical self-stimulation-like behavior for phasic VTA DA neuron firing.	
	VTA dopaminergic neurons	AAV-DIO-ChR2-EYFP	Blue light, 473nm 20Hz - Phasic firing	Synaptic alterations evidenced of enhanced non-GluR2 subunit insertion into DA VTA neurons after single injections of cocaine, morphine, or nicotine is mimicked by phasic ChR2 stimulation of VTA DA neurons.	Brown, et al. *PLoS One*, 2011

Figure 3. Optogenetic approaches in the reward circuitry.

effects of enhancing dopamine receptor signaling in NAc neurons) to elicit the CPP rewarding response. In addition, we were able to identify a molecular correlate associated with enhancing D_1^+ or D_2^+ MSN activity, which we discuss below.

As noted, the MSNs comprise 95% of all NAc and dStr neurons with the 5% of remaining neurons being interneurons. The tonically active cholinergic (also known as ChAT-choline acetyltranferase) interneurons make up ~1% of total NAc and dStr neurons and are the only known source of acetylcholine to striatum.[43] Because pharmacological and ablative studies demonstrate conflicting roles for cholinergic NAc interneurons in reward behaviors,[44,45] Witten *et al.* used optogenetics to examine a selective role for cholinergic interneurons after cocaine exposure and cocaine CPP (Fig. 3).[46] Using ChAT-Cre

mice and AAV-DIO-eNpHR3.0-EYFP virus, they demonstrated that acute cocaine exposure enhanced ChAT interneuron activity in the NAc, and optogenetic suppression of these interneurons suppressed cocaine CPP. Interestingly, silencing the ChAT interneurons increased activity in 76% and inhibited 24% of NAc neurons with a similar opposite response observed after ChR2 activation of the ChAT interneurons. Future work is needed to determine which population of NAc neurons are activated or inactivated by cocaine's activation of ChAT interneurons, or whether it is the D_1^+ or D_2^+ MSNs that are causing the cocaine-induced firing patterns displayed in ChAT interneurons. Current optogenetic tools and fluorescent labeling of specific neuronal subtypes have improved the feasibility to perform such studies. In fact, Chuma *et al.* recently demonstrated that 75% of striatal ChAT neurons hyperpolarize in response to ChR2 activation of striatal MSNs.[47] Such studies can now incorporate drugs of abuse to determine their effect on NAc and dStr functional connectome.

Manipulating NAc glutamatergic afferents with optogenetics

The NAc receives glutamatergic inputs from multiple brain regions, including prefrontal cortex, amygdala, and hippocampus, and there is a large body of evidence addressing the complex ways in which drugs of abuse alter glutamatergic neurotransmission to the NAc.[48–50] Recently, Stuber and colleagues demonstrated an important role for basolateral amygdala (BLA) inputs into the NAc in mediating rewarding behaviors (Fig. 3). This group showed that mice will self-stimulate optical activity of BLA glutamatergic inputs to NAc, but not mPFC inputs, using AAV viruses expressing ChR2 under the CAMK2α promoter.[51] Furthermore, they showed that the self-stimulation of BLA inputs required D_1 receptor signaling, which is consistent with the role of D_1 MSNs in enhancing reward behaviors. Finally, they revealed that silencing BLA inputs to NAc using AAV-CAMK2α-eNpHR3.0-EYFP reduces behavioral responding for a natural reward, sucrose. Future work evaluating molecular alterations in the NAc after optical self-stimulation of inputs into NAc will be very useful, given that such findings may correlate to known maladaptations in molecular machinery elicited by drugs of abuse.

Optogenetic manipulation of phasic VTA dopaminergic firing

Dopamine (DA) signaling from the VTA is well known for its critical role in regulating rewarding behaviors.[52,53] The switch in VTA DA cell firing from tonic firing (i.e., spontaneously occurring baseline spike activity that is driven by pacemaker-like membrane currents of DA neurons) to phasic firing (i.e., transient burst spike firing pattern triggering high amplitude DA release) correlates with unexpected rewards or sensory signals predicting such rewards.[53] Tsai and colleagues used tyrosine hydroxylase (which synthesizes DA)-Cre mice combined with AAV-DIO-ChR2-EYFP and demonstrated that mice display enhanced place preference for phasic ChR2 VTA DA stimulation and enhanced DA release (Fig. 3).[54]

More recently, using a food-seeking operant task, researchers demonstrated that phasic activation of VTA DA neurons, using the same genetic targeting described above, was sufficient to induce positive reinforcement during the food-seeking operant behavior and reactivate previously extinguished food-seeking behaviors in the absence of external cues (Fig. 3). Interestingly, mice did not display an optical self-stimulation-like behavior for phasic DA firing, which implies that DA phasic firing alone may not be sufficient for initiating a "liking" or a "wanting" signal.[55] Furthermore, it was recently shown that DA neurons can release glutamate[56,57] and that phasic VTA DA cell firing can mediate aversive cues, such as social avoidance after chronic social defeat stress,[58] thus illustrating the complex role of VTA DA neuron firing in behavioral encoding. Future studies using optogenetics to differentiate activity from different VTA DA neuronal populations (i.e., differentiating between those projecting to NAc and other brain regions) may help unravel their complex role in behavioral encoding.

Optogenetics to decipher structural plasticity

Drugs of abuse promote structural synaptic changes in NAc MSNs—in particular, dendritic spine alterations after cocaine exposure.[59] Our group is exploring the role of Rac1, a rho GTPase with known function in actin cytoskeleton dynamics. Using the photoactivatable Rac1 protein (Fig. 2) described above, we temporally activate Rac1 at critical time points, which thereby mimics cocaine exposure, to evaluate the role of Rac1 in cocaine induced spinogenesis and cocaine place preference.[37]

Other groups are using optogenetics to alter neuronal activity and then probe for functional alterations at synapses. Recently Brown et al. expressed ChR2 in VTA DA neurons using a dopamine transporter (DAT)-Cre line (Fig. 3). First, they showed that single exposure to drugs of abuse, including cocaine, morphine, and nicotine, enhanced glutamate receptor (GluR) rectification indexes (RI) in VTA DA neurons—a finding that is indicative of increased non-GluR2 (the calcium impermeable glutamate receptor subunit) synaptic activity—and increased non-GluR2 subunit insertion into the synapse of DA neurons. They then showed the enhanced GluR RIs in VTA DA neurons is mimicked by phasic stimulation of these neurons.[60] These data implicate a requirement for DA release to cause synaptic changes associated at least with the early stages of drug exposure; however, future work, combining chronic drugs of abuse with chronic optogenetic stimulation, may help to understand long-term plasticity alterations associated with addiction.

Because structural synaptic changes are a hallmark of exposure to drugs of abuse, more research studies will use these optogenetic approaches to study these phenomena. As noted above, DA neuronal firing promotes α-amino-3-hydroxy-5-methyl-4-isoxazolepropionic acid receptor (AMPAR) synaptic changes; and it will be interesting to see if optogenetic manipulation of VTA neurons can promote alterations in DA neuron cell size observed after chronic morphine treatment.[61,62] Furthermore, it will be interesting to see whether optogenetic manipulation of excitatory inputs from prefrontal cortex (PFC), basolateral amygdala (BLA), VTA, or ventral subiculum (VS) or dopaminergic inputs from VTA to MSNs in the NAc can alter the synaptic plasticity alterations observed after exposure to drugs of abuse.[59]

Probing genetics with optogenetics

Clearly, optogenetics has had a profound impact on neuroscience and neuroscientists, especially those studying the neurophysiology of intact circuits. However, those of us trying to decipher the molecular underpinnings of diseased states in neural circuits are also capitalizing on optogenetic techniques. As mentioned above we can use a photoactivatable intracellular signaling protein Rac1 to transiently activate it during cocaine exposure to probe its role in cocaine-induced spine plasticity and behavior.[37] We have also used optogenetics to selectively activate D_1^+ and D_2^+ MSNs, which enabled us to correlate a role for activity in these two MSNs to loss-of-function brain-derived neurotrophic factor (BDNF) signaling in cocaine reward.[25] Furthermore, we show that activating D_1^+ MSNs alters downstream effectors of BDNF signaling,[25] and have identified other molecular changes in the NAc after cell type–specific activation that play a role in mediating rewarding properties of drugs of abuse (unpublished data). In addition, we can use optogenetics to manipulate inputs into the NAc and probe for molecular changes in D_1^+ and D_2^+ MSNs, and correlate such findings to changes observed after exposure to drugs of abuse. In particular we can evaluate expression of ΔFosB, a long lasting iso-

form of the FosB gene that is induced robustly and persistently primarily in D_1^+ MSNs after chronic exposure to drugs of abuse and natural reward,[63] in each MSN subtype after chronic enhancement of VTA, PFC, BLA, and VS firing into the NAc. Finally, Koo and colleagues are using optogenetics to reverse the blunted morphine place preference response observed after increasing BDNF levels in the VTA, by enhancing firing of VTA inputs into the NAc using ChR2.[64] Other future studies using optogenetics to rescue behavioral phenotypes induced by molecular manipulations are likely on the horizon.

Such experiments above illustrate that optogenetics can be used by the molecular neuroscience field to probe meaningful genetic changes after altering functional or molecular activity with light or in the latter case to rescue a molecular induced phenotype. Nonetheless, tremendous advances are still needed to make this technology applicable to the general molecular neuroscience field and to molecular biologists studying other systems. First, we need to generate additional novel photoactivatable molecules involved in cellular and transcriptional functions. The currently available molecules are OptoXRs, photoactivatable adenylyl and guanylyl cyclase, and LOV domain fusion molecules (Fig. 2).[30,32–36,38] The LOV domain could plausibly be used to tether mammalian signaling genes and transcription factors then activate them with exposure to light. Photoactivatable transcription factors are very enticing given that drugs of abuse produce persistent alterations in transcriptional and epigenetic machinery;[65] hence, the ability to target transcriptional mechanisms on a very precise temporal scale will help pinpoint the timing and necessity of specific molecules in mediating the effects of drugs of abuse, and may help to generate newer therapeutics relevant to the timing of the addiction phase in human patients.

Conclusions

Optogenetic technologies have made a huge impact on researchers studying rewarding phenomena. In particular, the field has made advances toward understanding the role of novel circuitry and their cell types in rewarding behaviors and maladaptive responses to drugs of abuse using these light-activated molecules. Although most optogenetic studies involve acute optical stimulation protocols (usually within one to two days), it would be very informative and interesting to look at the chronic effects of manipulating neuronal activity patterns or chronic activation of molecular machinery and correlate them with maladaptive adaptations caused by drugs of abuse. Optogenetics will continue to increase our understanding of circuit and cell type functions in the CNS and thus will surely be an important experimental tool for developing better therapeutic targets for diseases of the central nervous system.

Acknowledgments

The author thanks D.M.D. at SUNY Buffalo for advice and discussion during manuscript preparation. The author thanks D.M.D. at SUNY at Buffalo, and E.J.N. and J-W.K. at MSSM for allowing inclusion of their unpublished work in this manuscript.

Conflicts of interest

The author declares no conflicts of interest.

References

1. Nestler, E.J. 2005. Is there a common molecular pathway for addiction? *Nat. Neurosci.* **8:** 1445–1449.
2. Nestler, E.J. & W.A. Carlezon, Jr. 2006. The mesolimbic dopamine reward circuit in depression. *Biol. Psychiatry* **59:** 1151–1159.
3. O'Donnell, P. 2011. Adolescent onset of cortical disinhibition in schizophrenia: insights from animal models. *Schizophr. Bull.* **37:** 484–492.
4. Livet, J. *et al.* 2007. Transgenic strategies for combinatorial expression of fluorescent proteins in the nervous system. *Nature* **450:** 56–62.
5. Gong, S. *et al.* 2003. A gene expression atlas of the central nervous system based on bacterial artificial chromosomes. *Nature* **425:** 917–925.
6. Gong, S. *et al.* 2007. Targeting Cre recombinase to specific neuron populations with bacterial artificial chromosome constructs. *J. Neurosci.* **27:** 9817–9823.
7. Yizhar, O. *et al.* 2011. Optogenetics in neural systems. *Neuron* **71:** 9–34.
8. Fenno, L., O. Yizhar & K. Deisseroth. 2011. The development and application of optogenetics. *Annu. Rev. Neurosci.* **34:** 389–412.
9. Nestler, E.J. 2001. Molecular basis of long-term plasticity underlying addiction. *Nat. Rev. Neurosci.* **2:** 119–128.
10. Badiani, A. *et al.* 1998. Amphetamine-induced behavior, dopamine release, and c-fos mRNA expression: modulation by environmental novelty. *J. Neurosci.* **18:** 10579–10593.
11. Koya, E. *et al.* 2009. Targeted disruption of cocaine-activated nucleus accumbens neurons prevents context-specific sensitization. *Nat. Neurosci.* **12:** 1069–1073.
12. Zemelman, B.V. *et al.* 2002. Selective photostimulation of genetically chARGed neurons. *Neuron* **33:** 15–22.
13. Volgraf, M. *et al.* 2006. Allosteric control of an ionotropic glutamate receptor with an optical switch. *Nat. Chem. Biol.* **2:** 47–52.
14. Banghart, M. *et al.* 2004. Light-activated ion channels for remote control of neuronal firing. *Nat. Neurosci.* **7:** 1381–1386.
15. Boyden, E.S. *et al.* 2005. Millisecond-timescale, genetically targeted optical control of neural activity. *Nat. Neurosci.* **8:** 1263–1268.
16. Nagel, G. *et al.* 2005. Light activation of channelrhodopsin-2 in excitable cells of Caenorhabditis elegans triggers rapid behavioral responses. *Curr. Biol.* **15:** 2279–2284.
17. Zhang, F. *et al.* 2007. Multimodal fast optical interrogation of neural circuitry. *Nature* **446:** 633–639.
18. Gradinaru, V. *et al.* 2007. Targeting and readout strategies for fast optical neural control in vitro and in vivo. *J. Neurosci.* **27:** 14231–14238.
19. Arenkiel, B.R. *et al.* 2007. In vivo light-induced activation of neural circuitry in transgenic mice expressing channelrhodopsin-2. *Neuron* **54:** 205–218.
20. Wang, H. *et al.* 2007. High-speed mapping of synaptic connectivity using photostimulation in Channelrhodopsin-2 transgenic mice. *Proc. Natl. Acad. Sci. USA.* **104:** 8143–8148.
21. Gradinaru, V. *et al.* 2009. Optical deconstruction of parkinsonian neural circuitry. *Science* **324:** 354–359.
22. Kravitz, A.V. *et al.* 2010. Regulation of parkinsonian motor behaviours by optogenetic control of basal ganglia circuitry. *Nature* **466:** 622–626.
23. Tye, K.M. *et al.* 2011. Amygdala circuitry mediating reversible and bidirectional control of anxiety. *Nature* **471:** 358–362.
24. Covington, H.E., 3rd, *et al.* 2010. Antidepressant effect of optogenetic stimulation of the medial prefrontal cortex. *J. Neurosci.* **30:** 16082–16090.
25. Lobo, M.K. *et al.* 2010. Cell type-specific loss of BDNF signaling mimics optogenetic control of cocaine reward. *Science* **330:** 385–390.
26. Zhang, F. *et al.* 2008. Red-shifted optogenetic excitation: a tool for fast neural control derived from Volvox carteri. *Nat. Neurosci.* **11:** 631–633.
27. Chow, B.Y. *et al.* 2010. High-performance genetically targetable optical neural silencing by light-driven proton pumps. *Nature* **463:** 98–102.
28. Gradinaru, V. *et al.* 2010. Molecular and cellular approaches for diversifying and extending optogenetics. *Cell* **141:** 154–165.
29. Han, X. *et al.* 2011. A high-light sensitivity optical neural silencer: development and application to optogenetic control of non-human primate cortex. *Front. Syst. Neurosci.* **5:** 1–8.
30. Airan, R.D. *et al.* 2009. Temporally precise in vivo control of intracellular signalling. *Nature* **458:** 1025–1029.
31. Oh, E. *et al.* 2010. Substitution of 5-HT1A receptor signaling by a light-activated G protein-coupled receptor. *J. Biol. Chem.* **285:** 30825–30836.
32. Ryu, M.H. *et al.* 2010. Natural and engineered photoactivated nucleotidyl cyclases for optogenetic applications. *J. Biol. Chem.* **285:** 41501–41508.
33. Stierl, M. *et al.* 2011. Light modulation of cellular cAMP by a small bacterial photoactivated adenylyl cyclase, bPAC, of the soil bacterium Beggiatoa. *J. Biol. Chem.* **286:** 1181–1188.
34. Hahn, K.M. & B. Kuhlman. 2010. Hold me tightly LOV. *Nat. Methods* **7:** 595, 597.

35. Wu, Y.I. *et al.* 2011. A genetically encoded photoactivatable Rac controls the motility of living cells. *Nature* **461:** 104–108.

36. Wu, Y.I. *et al.* 2009. Spatiotemporal control of small GTPases with light using the LOV domain. *Methods Enzymol.* **497:** 393–407.

37. Dietz, D.M. *et al.* 2011. The role of Rac1 in mediating cocaine induced structural plasticity. Neuroscience Meeting Planner. Washington, DC: Society for Neuroscience. Online **2011:** Abstract no. 909.922.

38. Strickland, D., K. Moffat & T.R. Sosnick. 2008. Light-activated DNA binding in a designed allosteric protein. *Proc. Natl. Acad. Sci. USA* **105:** 10709–10714.

39. Lobo, M.K. 2009. Molecular profiling of striatonigral and striatopallidal medium spiny neurons past, present, and future. *Int. Rev. Neurobiol.* **89:** 1–35.

40. Gerfen, C.R. *et al.* 1990. D1 and D2 dopamine receptor-regulated gene expression of striatonigral and striatopallidal neurons. *Science* **250:** 1429–1432.

41. Gerfen, C.R. 1992. The neostriatal mosaic: multiple levels of compartmental organization in the basal ganglia. *Annu. Rev. Neurosci.* **15:** 285–320.

42. Albin, R.L., A.B. Young & J.B. Penney. 1989. The functional anatomy of basal ganglia disorders. *Trends Neurosci.* **12:** 366–375.

43. Zhou, F.M., C.J. Wilson & J.A. Dani. 2002. Cholinergic interneuron characteristics and nicotinic properties in the striatum. *J. Neurobiol.* **53:** 590–605.

44. Hikida, T. *et al.* 2001. Increased sensitivity to cocaine by cholinergic cell ablation in nucleus accumbens. *Proc. Natl. Acad. Sci. USA* **98:** 13351–13354.

45. Pontieri, F.E. *et al.* 1996. Effects of nicotine on the nucleus accumbens and similarity to those of addictive drugs. *Nature* **382:** 255–257.

46. Witten, I.B. *et al.* 2010. Cholinergic interneurons control local circuit activity and cocaine conditioning. *Science* **330:** 1677–1681.

47. Chuhma, N. *et al.* 2011. Functional connectome of the striatal medium spiny neuron. *J. Neurosci.* **31:** 1183–1192.

48. Beurrier, C. & R.C. Malenka. 2002. Enhanced inhibition of synaptic transmission by dopamine in the nucleus accumbens during behavioral sensitization to cocaine. *J. Neurosci.* **22:** 5817–5822.

49. Kalivas, P.W. 2009. The glutamate homeostasis hypothesis of addiction. *Nat. Rev. Neurosci.* **10:** 561–572.

50. Wolf, M.E. 2010. Regulation of AMPA receptor trafficking in the nucleus accumbens by dopamine and cocaine. *Neurotox. Res.* **18:** 393–409.

51. Stuber, G.D. *et al.* 2011. Excitatory transmission from the amygdala to nucleus accumbens facilitates reward seeking. *Nature* **475:** 377–380.

52. Wise, R.A. 2004. Dopamine, learning and motivation. *Nat. Rev. Neurosci.* **5:** 483–494.

53. Schultz, W. 1997. Dopamine neurons and their role in reward mechanisms. *Curr. Opin. Neurobiol.* **7:** 191–197.

54. Tsai, H.C. *et al.* 2009. Phasic firing in dopaminergic neurons is sufficient for behavioral conditioning. *Science* **324:** 1080–1084.

55. Adamantidis, A.R. *et al.* 2011. Optogenetic interrogation of dopaminergic modulation of the multiple phases of reward-seeking behavior. *J. Neurosci.* **31:** 10829–10835.

56. Hnasko, T.S. *et al.* 2010. Vesicular glutamate transport promotes dopamine storage and glutamate corelease in vivo. *Neuron.* **65:** 643–656.

57. Stuber, G.D. *et al.* 2010. Dopaminergic terminals in the nucleus accumbens but not the dorsal striatum corelease glutamate. *J. Neurosci.* **30:** 8229–8233.

58. Cao, J.L. *et al.* 2010. Mesolimbic dopamine neurons in the brain reward circuit mediate susceptibility to social defeat and antidepressant action. *J. Neurosci.* **30:** 16453–16458.

59. Dietz, D.M. *et al.* 2009. Molecular mechanisms of psychostimulant-induced structural plasticity. *Pharmacopsychiatry* **42**(Suppl. 1): S69–S78.

60. Brown, M.T. *et al.* 2010. Drug-driven AMPA receptor redistribution mimicked by selective dopamine neuron stimulation. *PLoS. One.* **5:** e15870.

61. Sklair-Tavron, L. *et al.* 1996. Chronic morphine induces visible changes in the morphology of mesolimbic dopamine neurons. *Proc. Natl. Acad. Sci. USA* **93:** 11202–11207.

62. Russo, S.J. *et al.* 2007. IRS2-Akt pathway in midbrain dopamine neurons regulates behavioral and cellular responses to opiates. *Nat. Neurosci.* **10:** 93–99.

63. Nestler, E.J. 2008. Review. Transcriptional mechanisms of addiction: role of DeltaFosB. *Philos. Trans. R. Soc. Lond. B. Biol. Sci.* **363:** 3245–3255.

64. Koo, J.W. *et al.* 2011. Role of BDNF in the VTA in regulating molecular and behavioral responses to morphine. Neuroscience Meeting Planner. Washington, DC: Society for Neuroscience. Online **2011:** Abstract no. 167.101.

65. Maze, I. & E.J. Nestler. 2011. The epigenetic landscape of addiction. *Ann. N.Y. Acad. Sci.* **1216:** 99–113.

Ann. N.Y. Acad. Sci. ISSN 0077-8923

ANNALS OF THE NEW YORK ACADEMY OF SCIENCES

Issue: *Blavatnik Awards for Young Scientists*

Searching for new physics at the frontiers with lattice quantum chromodynamics

Ruth S. Van de Water

Physics Department, Brookhaven National Laboratory, Upton, New York

Address for correspondence: Ruth S. Van de Water, Physics Department, Brookhaven National Laboratory, Upton, NY 11973. ruthv@bnl.gov

Numerical lattice–quantum chromodynamics (QCD) simulations, when combined with experimental measurements, allow the determination of fundamental parameters of the particle-physics Standard Model and enable searches for physics beyond-the-Standard Model. We present the current status of lattice–QCD weak matrix element calculations needed to obtain the elements and phase of the Cabibbo–Kobayashi–Maskawa (CKM) matrix and to test the Standard Model in the quark-flavor sector. We then discuss evidence that may hint at the presence of new physics beyond the Standard Model CKM framework. Finally, we discuss two opportunities where we expect lattice QCD to play a pivotal role in searching for, and possibly discovery of, new physics at upcoming high-intensity experiments: rare $K \rightarrow \pi\nu\bar{\nu}$ decays and the muon anomalous magnetic moment. The next several years may witness the discovery of new elementary particles at the Large Hadron Collider (LHC). The interplay between lattice QCD, high-energy experiments at the LHC, and high-intensity experiments will be needed to determine the underlying structure of whatever physics beyond-the-Standard Model is realized in nature.

Keywords: lattice QCD; CKM matrix; Standard Model

Lattice quantum chromodynamics (QCD) and the search for new physics

One of the foremost goals of elementary particle physics is to test the Standard Model and to search for new physics beyond the Standard Model. The Standard Model of particle physics describes all observed elementary particles and their electromagnetic, weak, and strong interactions. It has proven to be a remarkably successful theory, and accounts for the outcome of every high-energy particle physics experiment to date. Nevertheless, there are numerous reasons to believe that the Standard Model is only a low-energy effective description of a more fundamental theory. For example, the Standard Model does not account for neutrino masses, the amount of dark matter and dark energy in the universe, or the abundance of matter over antimatter. In the quark-flavor sector, it provides no explanation for why there are three generations of quarks, or what generates the hierarchy of quark masses and Cabibbo–Kobayashi–Maskawa (CKM) mixing matrix elements.

Experimental particle physicists rely on two complementary approaches to search for physics beyond the Standard Model. The first is to try to create new particles directly at the *energy frontier*: this requires machines such as the Large Hadron Collider (LHC), which is now colliding protons at a center-of-mass energy of 7 TeV. The second approach is to make precise experimental measurements at the *intensity frontier* and look for discrepancies with the Standard Model. Theoretical arguments based on naturalness and fine-tuning suggest that new physics is preferred near the TeV scale, so we expect to discover new particles at the LHC. Precision measurements in the quark-flavor sector, however, rule out the possibility of large TeV-scale new-physics contributions to flavor-changing neutral currents. Thus whatever TeV-scale new physics is realized in nature must have a highly nongeneric flavor structure; for example, the new phases may be aligned with

doi: 10.1111/j.1749-6632.2012.06542.x

Ann. N.Y. Acad. Sci. 1260 (2012) 34–44 © 2012 New York Academy of Sciences.

the Standard Model such as in Minimal Flavor Violation[1] or the new couplings may be unnaturally tiny. This is called the *new-physics flavor problem*. Experiments at both the energy and intensity frontiers are necessary to resolve this puzzle. If new particles are discovered at the LHC, precision flavor measurements will be needed to extract their couplings and determine the underlying structure of the theory. If new physics unfortunately turns out to live above the TeV scale, indirect searches at the intensity frontier will be our only probe of physics beyond-the-Standard Model.

In the forthcoming decade many new high-intensity experiments will begin running to measure quantities as diverse as neutrino masses and mixing parameters, rare kaon and *B*-meson decays, and the muon anomalous magnetic moment. In order to maximize the scientific output of these upcoming experiments and reveal potential new physics signals, it is crucial to provide reliable and precise theoretical determinations of the Standard Model predictions. Many of these predictions require determinations of hadronic parameters such as decay constants, form factors, and meson-mixing matrix elements. Because these parameters encode nonperturbative QCD effects, they can only be calculated from first principles with all systematic uncertainties under control using numerical lattice QCD.

Introduction to lattice QCD

QCD governs the strong nuclear interactions. It determines how quarks and gluons, the fundamental constituents of ordinary matter, bind together to form composite particles (hadrons) such as protons and neutrons, and how these particles in turn interact to form atomic nuclei. At high energies the QCD coupling is small, and one can calculate QCD processes with perturbation theory, that is, a series expansion in the coupling constant α_S. At low energies, the QCD coupling becomes of $\mathcal{O}(1)$ or larger and the perturbative expansion breaks down. In this regime, quarks and gluons become confined into hadrons and one must solve QCD nonperturbatively, that is, to all orders in α_S. Currently the only available method for including the effect of confining quarks into hadrons from QCD first principles is with lattice QCD.

Lattice gauge theory is a general tool for solving nonperturbative quantum field theories (see Refs. 2–4 for pedagogical introductions to lattice methods). Lattice–QCD calculations formulate QCD on a discrete Euclidean spacetime lattice, thereby transforming the infinite-dimensional Quantum Field Theory path integral into a finite-dimensional integral that can be solved numerically with Monte Carlo methods and importance sampling. In practice, lattice–QCD simulations are computationally intensive, and require the use of the world's most powerful supercomputers.

Lattice QCD provides *ab initio* QCD calculations of many quantities that cannot be obtained via experiment such as quark masses and weak-interaction matrix elements. Over the past decade, increased computing power and better algorithms have enabled realistic QCD calculations that include the effects of the dynamical *u*, *d*, and *s* quarks in the vacuum polarization. This major breakthrough has led to substantial progress in the computation of quantities needed to interpret experimental results in the areas of particle physics, nuclear physics, and even astrophysics. Lattice–QCD calculations now reproduce experimental results for a wide variety of hadron properties, as shown in Figure 1. For example, the most precise determination of the strong coupling constant α_S is obtained by computing 28 short-distance lattice quantities such as Wilson loops and then fitting the β-function to the next-to-next-to-leading order perturbative-QCD expression.[5,6] Lattice–QCD calculations also correctly predicted the mass of the B_c meson,[7,8] the leptonic decay constants f_D and f_{D_s},[9,10] and the $D \to K \ell \nu$ semileptonic form factor.[11,12] These successful predictions and postdictions demonstrate that the systematic uncertainties in lattice–QCD calculations are under control, and that lattice–QCD results can reliably be used to test the Standard Model and search for new physics. Further, lattice–QCD uncertainties will be steadily reduced over the next several years given the expected increase in computing power.

Status of the CKM matrix and CKM unitarity triangle

Most Standard Model extensions contain new *CP*-violating phases and new quark-flavor changing interactions that could give rise to observable deviations from Standard Model predictions given sufficient experimental and theoretical precision. In the following sections we present the current status of lattice–QCD calculations needed to test the

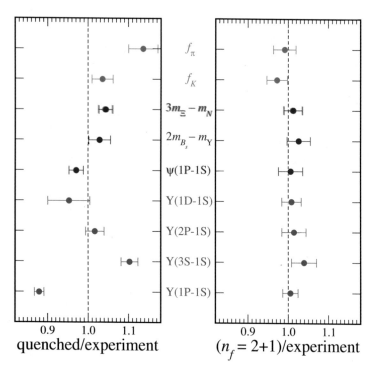

Figure 1. Ratio of lattice–QCD results over experimental measurements for a selection of weak-matrix elements and meson and baryon mass splittings. The left-hand panel corresponds to "quenched" simulations in which the effects of dynamical quarks are omitted, while the right-hand panel corresponds to realistic "2+1 flavor" simulations that include the effects of the dynamical u, d, and s quarks. Figure updated from Ref. 13 to reflect more recent lattice calculations.

Standard Model in the quark-flavor sector, and then briefly discuss an exciting recent development that may hint at the presence of physics beyond-the-Standard Model.

The Standard Model CKM framework

The CKM quark-mixing matrix parameterizes the mixing between quark flavors under weak interactions:[14,15]

$$V_{\mathrm{CKM}} = \begin{pmatrix} V_{ud} & V_{us} & V_{ub} \\ V_{cd} & V_{cs} & V_{cb} \\ V_{td} & V_{ts} & V_{tb} \end{pmatrix}. \quad (1)$$

The elements of the CKM matrix are fundamental parameters of the Standard Model; thus they are of interest to understand the Standard Model's flavor structure. Further, the CKM matrix elements are parametric inputs to Standard Model predictions for many flavor-changing processes such as neutral kaon mixing and $K \to \pi\nu\bar{\nu}$ decay, and must be known precisely to search for new physics in these channels.

Within the Standard Model, the CKM matrix is unitary ($V_{\mathrm{CKM}}^{\dagger} V_{\mathrm{CKM}} = \mathbb{1}$). One can therefore expand the elements of the CKM matrix in powers of $\lambda \equiv |V_{us}| \sim 0.22$:

$$V_{\mathrm{CKM}}$$
$$= \begin{pmatrix} 1 - \frac{1}{2}\lambda^2 & \lambda & A\lambda^3(\bar{\rho} - i\bar{\eta}) \\ -\lambda & 1 - \frac{1}{2}\lambda^2 & A\lambda^2 \\ A\lambda^3(1 - \bar{\rho} - i\bar{\eta}) & -A\lambda^2 & 1 \end{pmatrix}$$
$$+ \mathcal{O}(\lambda^4). \quad (2)$$

The Wolfenstein parameterization shown above[16] makes the hierarchy of sizes of the different matrix elements explicit. Further, unitarity ensures that there are relationships among the elements of the CKM matrix such as

$$V_{ud}V_{ub}^* + V_{cd}V_{cb}^* + V_{td}V_{tb}^* = 0. \quad (3)$$

This equation can be expressed as a complex triangle in the $\bar{\rho}$–$\bar{\eta}$ plane known as the CKM unitarity triangle, shown in Figure 2. A standard way of searching for new physics in the quark-flavor sector is by

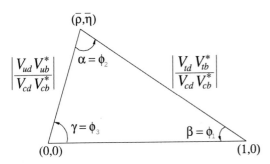

$$\begin{pmatrix}
\underset{\substack{V_{ud} \\ \pi \to \ell\nu}}{} & \underset{\substack{V_{us} \\ K \to \ell\nu \\ K \to \pi\ell\nu}}{} & \underset{\substack{V_{ub} \\ B \to \ell\nu \\ B \to \pi\ell\nu \\ B \to \tau\nu}}{} \\[2em]
\underset{\substack{V_{cd} \\ D \to \ell\nu \\ D \to \pi\ell\nu}}{} & \underset{\substack{V_{cs} \\ D_s \to \ell\nu \\ D \to K\ell\nu}}{} & \underset{\substack{V_{cb} \\ B \to D\ell\nu \\ B \to D^*\ell\nu}}{} \\[2em]
\underset{\substack{V_{td} \\ \langle B_d | \bar{B}_d \rangle}}{} & \underset{\substack{V_{ts} \\ \langle B_s | \bar{B}_s \rangle}}{} & \underset{V_{tb}}{}
\end{pmatrix}$$

Figure 2. The CKM unitarity triangle. Figure from Ref. 17.

Figure 3. "Gold-plated" processes in lattice–QCD simulations that can be used to obtain each CKM matrix element. Neutral $K^0 - \bar{K}^0$ mixing is also a gold-plated process, and can be used to obtain the phase of the CKM matrix $(\bar{\rho}, \bar{\eta})$.

overconstraining the angles and sides of the CKM unitarity triangle. New quark-flavor–changing interactions or *CP*-violating phases would manifest themselves as apparent inconsistencies between measurements of the apex $(\bar{\rho}, \bar{\eta})$ that are predicted to be the same within the Standard Model CKM framework.

Lattice–QCD inputs

The CKM matrix elements and phase are parametric inputs to the Standard Model, so they cannot be calculated from first principles, and must be obtained from experimental measurements. In practice, however, this procedure is complicated by the fact that one cannot observe free quarks in nature: to accurately describe weak interactions involving quarks, one must include the nonperturbative QCD effects of confining quarks into hadrons. Typically these complex dynamics are absorbed into quantities such as decay constants, form factors, and bag parameters, that must be computed numerically in lattice QCD. Precise lattice–QCD calculations of hadronic weak matrix elements are needed to interpret many experimental flavor physics results as determinations of CKM matrix elements and tests of the Standard Model; thus they are crucial for the success of the worldwide experimental flavor-physics program.

"Gold-plated" lattice processes allow the determination of all CKM matrix elements except $|V_{tb}|$, as shown in Figure 3. These are simple processes with a single hadron in the initial state and at most one hadron in the final state, where both hadrons are stable (or at least narrow and far from threshold). Because they are easiest to compute numerically with standard lattice methods, they are among the most well-studied lattice–QCD quantities. Realistic lattice–QCD calculations that include the effects of

the dynamical *u*, *d*, and *s* quarks are now available for all hadronic matrix elements listed in Figure 3. Further, for most of these quantities, there are at least two calculations by different collaborations that provide valuable cross-checks. To maximize the impact of lattice–QCD calculations on new-physics searches, Laiho *et al.* have compiled three-flavor averages of lattice–QCD weak matrix elements needed for CKM phenomenology.[18] Table 1 presents a summary of their main results updated to reflect the 2011 summer conferences.

Lattice–QCD calculations currently allow determinations of $|V_{us}|$, $|V_{cd}|$, $|V_{cs}|$, $|V_{ub}|$, and $|V_{cb}|$ that are competitive with the world's best.[24,29,30,36,39] Given these results, one can test the unitarity of the first and second rows of the CKM matrix:

$$|V_{ud}|^2 + |V_{us}|^2 + |V_{ub}|^2 = 1 \qquad \text{(first row)} \quad (4)$$

$$|V_{cd}|^2 + |V_{cs}|^2 + |V_{cb}|^2 = 1 \qquad \text{(second row)} \quad (5)$$

Deviations from these relationships would indicate the presence of physics beyond-the-Standard Model. Current lattice–QCD and experimental results are consistent with first-row unitarity at the subpercent level (see Fig. 4)[43] and with second-row unitarity at the percent-level.[29] This stringent test of first-row unitarity was enabled by precise lattice–QCD determinations of the ratio of leptonic decay constants f_K/f_π and of the $K \to \pi\ell\nu$ semileptonic form factor $f_+^{K\pi}(0)$, and places a constraint on the effective scale of generic new-physics contributions to Equation (4) be above $\mathcal{O}(10\text{TeV})$.[43]

Table 1. Averages of lattice–QCD inputs needed to obtain the elements and phase of the CKM matrix

$f_\pi = (129.8 \pm 1.5)\,\text{MeV}$[19–22]	$f_K = (156.1 \pm 1.1)\,\text{MeV}$[19–22]
$f_+^{K\pi}(0) = 0.9584 \pm 0.0044$[23,24]	$\hat{B}_K = 0.7643 \pm 0.0097$[20,22,25–27]
$f_D = (213.5 \pm 4.1)\,\text{MeV}$[19,28]	$f_{D_s} = (248.6 \pm 3.0)\,\text{MeV}$[19,28]
$f_+^{D\pi}(0) = 0.666 \pm 0.029$[29]	$f_+^{DK}(0) = 0.747 \pm 0.019$[30]
$f_B = (190.6 \pm 4.7)\,\text{MeV}$[28,31]	$f_{B_s} = (227.6 \pm 5.0)\,\text{MeV}$[31]
$f_{B_d}\sqrt{\hat{B}_{B_d}} = (227 \pm 19)\,\text{MeV}$[33,34]	$f_{B_s}\sqrt{\hat{B}_{B_s}} = (279 \pm 15)\,\text{MeV}$[33,34]
$\lvert V_{cb}\rvert_{\text{excl}} = (39.5 \pm 1.0) \times 10^{-3}$[35–37]	$\lvert V_{ub}\rvert_{\text{excl}} = (3.12 \pm 0.26) \times 10^{-3}$[38–41]

Note: Results are updated from Ref. 18 to reflect more recent lattice–QCD calculations (see www.latticeaverages.org and Ref. 42 for more details). For the elements $\lvert V_{ub}\rvert$ and $\lvert V_{cb}\rvert$, the lattice–QCD calculations of the form factors have already been combined with the experimentally measured branching fractions to obtain the CKM matrix elements.

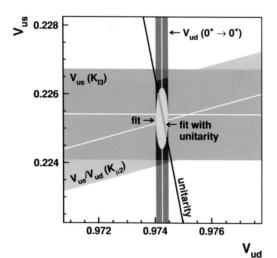

Figure 4. Test of first-row CKM unitarity. The yellow ellipse denotes the result of the fit to the quantities $\lvert V_{ud}\rvert$, $\lvert V_{us}\rvert$, and $\lvert V_{us}\rvert/\lvert V_{ud}\rvert$; the $\chi^2/\text{dof} = 0.014/1$ (P value = 91%). The brown line denotes the result of the fit to the same three quantities imposing the unitarity constraint in Equation (4); the $\chi^2/\text{dof} = 0.024/2$ (P value = 99%). Figure from Ref. 43.

A highlight of the worldwide lattice–QCD weak-matrix elements program is the calculation of the neutral kaon mixing parameter B_K, which is needed to interpret experimental measurements of indirect CP-violation in the kaon system (ϵ_K) as constraints on the apex of the CKM unitarity triangle. Because ϵ_K has been experimentally measured to subpercent accuracy, a similarly precise value of the hadronic parameter B_K is required to test the Standard Model and search for new physics. Until recently, the unitarity-triangle constraint from ϵ_K was limited by the approximately 20% uncertainty in lattice–QCD calculations of B_K;[44] therefore signifi-

cant theoretical and computational effort has been devoted to its improvement. In 2007 the RBC and UKQCD collaborations published the first realistic calculation of B_K that includes the effects of the dynamical u, d, and s quarks, but only at a single value of the lattice spacing.[45] In 2009, Aubin *et al.* followed this important result by publishing the first calculation of B_K with all sources of systematic uncertainty under control.[46] There are now several independent lattice–QCD results for B_K that are in good agreement,[20,22,25,27] and the total error in B_K is now below 2%.

A crack in the CKM paradigm?

Over the past decade, heavy-flavor experiments have been pouring out data needed to obtain the elements and phase of the CKM matrix.[47] In conjunction with precise lattice–QCD weak-matrix element calculations, the B-factories successfully established that the CKM paradigm of CP-violation describes experimental observations at the few-to-several-percent level; this led to the bestowal of the 2008 Nobel Prize in physics to Kobayashi and Maskawa. Recently, however, experimental measurements and lattice–QCD calculations have revealed an approximately 3σ tension in the CKM unitarity triangle,[48–50] as indicated by a low P value for the global fit to all constraints on the apex ($\bar{\rho}, \bar{\eta}$) of the CKM unitarity triangle (see Fig. 5). The primary source of this disagreement with the Standard Model is the measurement of $B \to \tau\nu$ decay,[42] while other processes contribute less significant tensions. Improved measurements of the $B \to \tau\nu$ branching fraction by the Belle II and superB experiments,[51] in conjunction with improved lattice–QCD calculations of the B-meson leptonic decay constant, will probe this tension to higher precision in the next few years, and may eventually establish the presence of new physics in the quark-flavor sector with higher significance.

Lattice QCD and the intensity frontier

The next several years will witness a renaissance of high-intensity experiments in the United States and elsewhere that will form the cornerstone of the U.S. experimental particle-physics program and be a principal component of the worldwide effort. In the following sections we discuss two opportunities where we expect lattice–QCD calculations to play a key role in searches for (and possibly discovery of) new physics at the intensity frontier.

Rare kaon decays

"Rare" quark-flavor–changing neutral current processes in which the leading-order Standard Model contribution is at the one-loop level are particularly good channels in which to search for new physics. This is because new heavy particles will typically enter the loops, giving rise to non–Standard Model contributions to the decay rates that may be significant. Therefore loop processes are sensitive to physics at much higher energy scales than can be probed in collider experiments, in some cases of $\mathcal{O}(1,000 - 10,000\,\text{TeV})$.[52] Further, new-physics contributions to rare processes may be easier to observe above the suppressed Standard Model background.

The rare kaon decays $K^+ \to \pi^+ \nu \bar{\nu}$ and $K_L^0 \to \pi^0 \nu \bar{\nu}$, in particular, are "golden modes" with great new physics discovery potential. Because the hadronic uncertainties are under good theoretical control, the Standard Model branching ratios are known to a precision unmatched by any other quark-flavor–changing neutral current process. Despite this fact, however, the branching ratios are still only known to ~10% due to the parametric uncertainty in $|V_{cb}|^4$.[36,53] The lowest-order Standard Model contributions to $K \to \pi \nu \bar{\nu}$ are from one-loop electroweak penguin diagrams, so this channel is sensitive to many new-physics scenarios such as Little-Higgs models, warped extra dimensions, and a fourth generation.[54,55] Further, spectacular deviations from the Standard Model predictions are possible in many of these theories, as shown in Figure 6.

The upcoming experiment NA62 at CERN will measure the $K^+ \to \pi^+ \nu \bar{\nu}$ branching fraction,[57]

while KOTO in Japan expects to make the first observation of $K_L^0 \to \pi^0 \nu \bar{\nu}$ decay.[58] Ultimately, the proposed experiment ORKA at Fermilab along with its successors at the Project X accelerator complex aim to collect $\mathcal{O}(1,000)$ events or more in both channels,[59] leading to few-percent errors in the branching ratios. To interpret the results of these precise experimental measurements as tests of the Standard Model, however, a reduction in the uncertainty on the CKM matrix element $|V_{cb}|$ will be needed. Improved lattice–QCD calculations of both the $B \to D\ell\nu$ and $B \to D^*\ell\nu$ form factors are underway,[36,60] which, in combination with increased computing resources over the next few years, should lead to a reduction in the error on $|V_{cb}|$ to below ~1.5% on the timescale of the upcoming experiments. With this precision, the error on the Standard Model branching fraction would be approximately 6% and commensurate with the target experimental uncertainties. Further, even a 30% deviation from the Standard Model, which is possible within many different new-physics scenarios, would lead to a 5σ discovery of physics beyond-the-Standard Model.

Muon anomalous magnetic moment

The anomalous magnetic moment of the muon is another excellent candidate for new-physics discovery. First, this is because it is has been measured experimentally to extremely high precision. And second, it is a sensitive probe of heavy mass scales in the TeV range and beyond.

Without quantum mechanics, the muon's magnetic moment would be $g = 2$. Radiative corrections modify this value, giving rise to the "anomaly"

$$a_\mu \equiv \frac{(g_\mu - 2)}{2}. \tag{6}$$

Because a_μ is solely due to quantum-mechanical effects, virtual heavy particles entering the loops can generate significant non–Standard Model contributions in many well-motivated new-physics scenarios such as Supersymmetry (SUSY)[61] and warped extra dimensions.[62] Further, a measurement of a_μ provides complementary information to the observations made at the LHC. For example, the contribution to a_μ in many supersymmetric scenarios is proportional to both $\tan(\beta)$ (the ratio of the up- to down-type Higgs Boson vacuum expectation values) and the sign of the μ-parameter (the coefficient of the term in the Lagrangian that mixes the

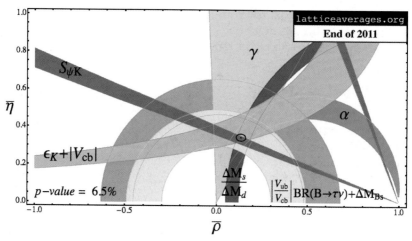

Figure 5. Global fit of the CKM unitarity triangle. The constraints labeled $\epsilon_K + |V_{cb}|$, $|V_{ub}/V_{cb}|$, $\Delta M_s/\Delta M_d$, and BR($B \to \tau \nu$) + ΔM_{B_s} all require inputs from lattice QCD. Figure updated from Ref. 18 to reflect more recent experimental measurements and lattice-QCD calculations; see www.latticeaverages.org and Ref. 42 for further details.

two Higgs doublets):[63]

$$a_\mu^{\text{SUSY}} \propto \left(\frac{m_\mu}{M_{\text{SUSY}}} \right)^2 \tan(\beta)\, \text{sign}(\mu)\,, \quad (7)$$

where M_{SUSY} is a generic SUSY mass scale. The parameters $\tan(\beta)$ and $\text{sign}(\mu)$ will be difficult to obtain precisely at the LHC because particle masses and cross sections are only indirectly sensitive to them (see, e.g., Ref. 64). Thus, if SUSY is discovered at the LHC, a measurement of a_μ can provide the additional information needed to pin down the parameters of the underlying theory realized in nature. In fact, the current measurement of a_μ already rules out many hypothesized new-physics models, as shown in Figure 7.

Experimentally, $(g-2)_\mu$ is known to 0.54 parts-per-million (ppm)[67] and disagrees with the Standard Model prediction by more than 3σ.[68] The quoted error in the Standard Model prediction for a_μ is 0.42 ppm. The QED and EW contributions to a_μ are known very precisely (to four loops and two loops in perturbation theory, respectively),[69,70] but the QCD contributions to a_μ are currently under less theoretical control[68,71] and their values must be taken with caution. Therefore both experimental and theoretical efforts are underway to investigate this discrepancy. The current measurement of a_μ is statistics limited, so the upcoming New $(g-2)$ Experiment aims to use the proton accelerator facility at Fermilab to reduce the experimental error in a_μ by approximately a factor of four to 0.14 ppm.[72] The

dominant errors in the Standard Model prediction for a_μ are from the hadronic vacuum polarization (0.36 ppm) and from hadronic light-by-light scattering (0.22 pm). The hadronic vacuum polarization contribution a_μ^{HVP} is currently obtained from combining experimental data for $e^+e^- \to$ hadrons with dispersion relations.[68] Errors in a_μ^{HVP} are expected to shrink by a factor of two on the timescale of the New $(g-2)$ Experiment, due to the analysis of larger data sets by KLOE and BABAR plus next-generation measurements by CMD and SND at Novosibirsk.[72] The current determination of the hadronic light-by-light contribution a_μ^{HLbL} is obtained from various models such as QCD in the large N_c limit, vector-meson dominance, and chiral perturbation theory.[71] The model results, however, are not consistent and the error in a_μ^{HLbL} is estimated from the spread of values. Improvement in understanding a_μ^{HLbL} is therefore critical to match the target experimental precision.

Because the hadronic light-by-light contribution arises due to nonperturbative QCD effects, the only way to calculate it from first principles is with lattice QCD. Research and development efforts to compute both a_μ^{HVP} and a_μ^{HLbL} are ongoing. The ETM Collaboration has developed an approach to reduce the chiral extrapolation error in a_μ^{HVP}.[73] There are also calculations of a_μ^{HVP} underway by Aubin and Blum[74] and by Boyle *et al.*[75] The calculation of the hadronic light-by-light contribution to the muon anomalous magnetic moment is significantly more

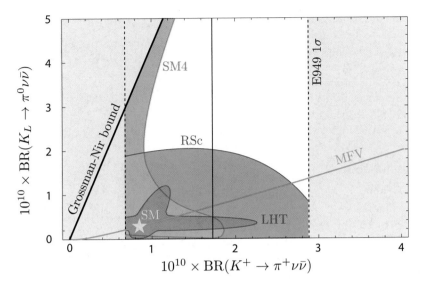

Figure 6. Correlations between the branching ratios $\mathrm{BR}(K_L \to \pi^0 \nu \bar{\nu})$ and $\mathrm{BR}(K^+ \to \pi^+ \nu \bar{\nu})$ in several new-physics models. Experimentally excluded regions are denoted by (light) gray bands. Figure from Ref. 56.

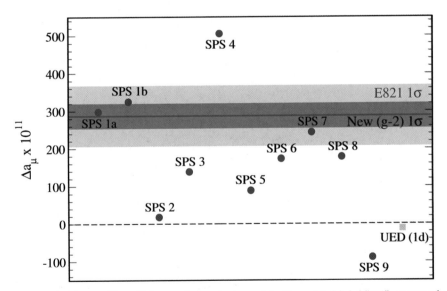

Figure 7. Predictions for $\Delta a_\mu \equiv a_\mu^{\mathrm{NP}} - a_\mu^{\mathrm{SM}}$ in several new-physics models. The points labeled "SPS" correspond to the SUSY benchmark points from the 2001 Snowmass workshop.[65] The outer blue (light gray) band shows the current experimental measurement of a_μ, while the inner magenta (dark gray) band shows the reduction in errors that should result from the New $(g-2)$ Experiment plus theoretical improvements, assuming a fixed central value. In this plot, SUSY predictions are from Ref. 61, while the universal extra dimension (UED) prediction is from Ref. 66.

challenging than that of the vacuum polarization, and lattice–QCD computations are still at an early stage. Cohen *et al.*[76] and the JLQCD Collaboration[77] are pursuing an alternative approach to compute the $\pi^0 \to \gamma\gamma$ form factor, which enters the dom-

inant contribution to a_μ^{HVP}. A promising method has been developed by Hayakawa *et al.* for obtaining a_μ^{HLbL} using QED + QCD lattice simulations that is simpler and cleaner than directly computing the correlation function of four currents,[78] but the precision goal for a_μ^{HLbL} is challenging, and both the expected increase in computing power and unforseen theoretical developments will likely be needed to reach it. Nevertheless, we expect that the necessary progress will be made and that the experimental measurement of the muon anomalous magnetic moment will in the next few years become an even more powerful tool to test the Standard Model and search for new physics.

Summary and outlook

Lattice–QCD provides a numerical tool for first-principles computations of hadronic matrix elements needed to interpret experimental measurements as determinations of Standard Model parameters and searches for new physics. Realistic lattice–QCD calculations including the effects of the dynamical u, d, and s quarks are now standard, and allow determinations of the elements of and phase of the CKM matrix, in some cases with percent-level precision. Lattice–QCD weak-matrix element calculations also enable tests of the Standard Model in the quark-flavor sector, and currently point to an approximately 3σ tension in the global CKM unitarity triangle fit that may indicate a non-Standard Model source of CP-violation. Both the experimental and theoretical inputs needed to probe this hint of new physics with higher significance are expected to improve in the next few years. Lattice–QCD calculations will soon be needed to interpret the results of upcoming intensity-frontier experiments such as those at the super-B factories, the CERN SPS and J-PARC, and Fermilab, and the lattice–QCD community is expanding its research program accordingly. The next decade should prove to be an exciting time in particle physics as we expect to discover TeV-scale new particles at the LHC. The interplay between measurements at the energy frontier and the intensity frontier, when combined with lattice–QCD matrix–element calculations, should help us to better understand and possibly resolve the new-physics flavor puzzle, and ultimately to pin down the parameters of whatever physics beyond-the-Standard Model is realized in nature.

Acknowledgments

I thank my collaborators Jack Laiho and Enrico Lunghi for fruitful discussions on these topics and essential contributions to some of the work described here.

Conflicts of interest

The author declares no conflicts of interest.

References

1. D'Ambrosio, G., G. Giudice, G. Isidori & A. Strumia. 2002. *Nucl. Phys.* **B645**: 155. arXiv:hep-ph/0207036 [hep-ph].
2. Creutz, M. 1985. *Quarks, Gluons and Lattices, Cambridge Monographs on Mathematical Physics.* Cambridge University Press.
3. Montvay, I. & G. Münster. 1997. *Quantum Fields on a Lattice, Cambridge Monographs on Mathematical Physics.* Cambridge University Press.
4. DeGrand, T. & C. DeTar. 2006. *Lattice Methods for Quantum Chromodynamics.* World Scientific Publishing Company.
5. Davies, C. *et al.* (HPQCD Collaboration). 2008. *Phys. Rev.* **D78**, 114507. arXiv:0807.1687 [hep-lat].
6. Bethke, S. 2009. *Eur. Phys. J.* **C64**: 689. arXiv:0908.1135 [hep-ph].
7. Allison, I.F. *et al.* (HPQCD Collaboration). 2005. *Phys. Rev. Lett.* **94**: 172001. arXiv:hep-lat/0411027 [hep-lat].
8. Abulencia, A. *et al.* (CDF Collaboration). 2006. *Phys. Rev. Lett.* **96**: 082002. arXiv:hep-ex/0505076 [hep-ex].
9. Aubin, C. *et al.* (Fermilab Lattice, MILC, and HPQCD Collaborations). 2005. *Phys. Rev. Lett.* **95**: 122002. arXiv:hep-lat/0506030 [hep-lat].
10. Artuso, M. *et al.* (CLEO Collaboration). 2005. *Phys. Rev. Lett.* **95**: 251801. arXiv:hep-ex/0508057 [hep-ex].
11. Aubin, C. *et al.* (Fermilab Lattice, MILC, and HPQCD Collaborations). 2005. *Phys. Rev. Lett.* **94**: 011601. arXiv:hep-ph/0408306 [hep-ph].
12. Abe, K. *et al.* (BELLE Collaboration). 2005. arXiv:hep-ex/0510003 [hep-ex].
13. Davies, C. *et al.* (HPQCD, UKQCD, MILC, and Fermilab Lattice Collaborations). 2004. *Phys. Rev. Lett.* **92**: 022001. arXiv:hep-lat/0304004 [hep-lat].
14. Cabibbo, N. 1963. *Phys. Rev. Lett.* **10**: 531.
15. Kobayashi, M. & T. Maskawa. 1973. *Prog. Theor. Phys.* **49**: 652.
16. Wolfenstein, L. 1983. *Phys. Rev. Lett.* **51**: 1945.
17. Nakamura, K. *et al.* (Particle Data Group). 2010. *J. Phys. G* **G37**: 075021.
18. Laiho, J., E. Lunghi, & R.S. Van de Water. 2010. *Phys. Rev.* **D81**: 034503. Available at: http://www.latticeaverages.org, arXiv:0910.2928 [hep-ph].
19. Davies, C. *et al.*, (HPQCD Collaboration). 2010. Update: precision D_s decay constant from full lattice QCD using very fine lattices. *Phys. Rev.* **D82**: 114504, [arXiv:1008.4018].
20. Laiho, J. & R. S. Van de Water. 2011. Pseudoscalar decay constants, light-quark masses, and B_K from mixed-action lattice QCD. *PoS LAT.* **2011**: 293. [arXiv:1112.4861].
21. Bazavov, A. *et al.* (MILC Collaboration). 2010. Results for light pseudoscalar mesons. *PoS LAT.* **2010**: 074. [arXiv:1012.0868].
22. Kelly, C. (RBC and UKQCD Collaboration). 2012. Continuum results for light Hadronic Quantities using domain wall Fermions with the Iwasaki and DSDR Gauge Actions. *PoS LAT.* **2011**: 285, [arXiv:1201.0706].

23. Lubicz, V., F. Mescia, S. Simula, & C. Tarantino, (ETM Collaboration). 2009. $K \to \pi \ell \nu$ semileptonic form factors from two-flavor lattice QCD. *Phys. Rev.* **D80**: 111502. [arXiv:0906.4728].

24. Boyle, P. *et al.* (RBC and UKQCD Collaboration). 2010. $K \to \pi$ form factors with reduced model dependence. *Eur. Phys. J.* **C69**: 159–167, [arXiv:1004.0886].

25. DŁurr, S. *et al.* (BMW Collaboration). 2011. Precision computation of the kaon bag parameter. *Phys. Lett.* **B705**: 477–481, [arXiv:1106.3230].

26. Gamiz, E. *et al.* (HPQCD, UKQCD Collaboration). 2006. Unquenched determination of the kaon parameter B_K from improved staggered fermions. *Phys. Rev.* **D73**: 114502. [hep-lat/0603023].

27. Bae, T. *et al.* (SWME Collaboration). Kaon B-parameter from improved staggered fermions in $N_f = 2 + 1$ QCD. arXiv: 1111.5698.

28. Bazavov, A. *et al.* (Fermilab Lattice and MILC Collaboration). *B*- and *D*-meson decay constants from three-flavor lattice QCD. arXiv: 1112.3051.

29. Na, H. *et al.* (HPQCD Collaboration). 2011. $D \to \pi, \ell \nu$ semileptonic decays, $|V_{cd}|$ and second row unitarity from lattice QCD. *Phys. Rev.* **D84**: 114505. [arXiv:1109.1501].

30. Na, H., C.T. Davies, E. Follana, *et al.* (HPQCD Collaboration). 2010. $D \to K$, $\ell \nu$ semileptonic decay scalar form factor and $|V_{cs}|$ from lattice QCD. *Phys. Rev.* **D82**: 114506. [arXiv:1008.4562].

31. Na, H. *et al.* (HPQCD Collaboration). 2012. The B and B_s Meson decay constants from lattice QCD. arXiv:1202.4914.

32. McNeile, C., C. Davies, E. Follana, *et al.* (HPQCD Collaboration). 2012. High-Precision f_{Bs} and HQET from Relativistic Lattice QCD. *Phys. Rev.* **D85**: 031503. [arXiv:1110.4510].

33. Bouchard, C. *et al.* (Fermilab Lattice and MILC Collaboration). Neutral *B* mixing from 2 + 1 flavor lattice-QCD: the Standard Model and beyond. arXiv:1112.5642.

34. Gamiz, E., C.T. Davies, G.P. Lepage, *et al.* (HPQCD Collaboration). 2009. Neutral *B* Meson mixing in unquenched lattice QCD. *Phys. Rev.* **D80**: 014503. [arXiv:0902.1815].

35. Okamoto, M. *et al.* (Fermilab Lattice and MILC Collaboration). 2005. Semileptonic $D \to \pi/K$ and $B \to \pi/D$ decays in 2+1 flavor lattice QCD. *Nucl. Phys. Proc. Suppl.* **140**: 461–463. [hep-lat/0409116].

36. Bailey, J.A. *et al.* (Fermilab Lattice and MILC Collaboration). 2010. $B \to D^* \ell \nu$ at zero recoil: an update. *PoS LAT.* **2010**: 311, [arXiv:1011.2166].

37. Asner, D. *et al.* (Heavy Flavor Averaging Group Collaboration). Averages of *b*-hadron, *c*-hadron, and τ-lepton Properties. arXiv:1010.1589.

38. Dalgic, E. *et al.* (HPQCD Collaboration). 2006. B meson semileptonic form-factors from unquenched lattice QCD. *Phys. Rev.* **D73**: 074502, [hep-lat/0601021].

39. Bailey, J.A. *et al.* (Fermilab Lattice and MILC Collaboration). 2009. The $B \to \pi \ell \nu$ semileptonic form factor from three-flavor lattice QCD: a model-independent determination of $|V_{ub}|$. *Phys. Rev.* **D79**: 054507. [arXiv:0811.3640].

40. del Amo Sanchez, P. *et al.* (BABAR Collaboration). 2011. Study of $B \to \pi \ell \nu$ and $B \to \rho \ell \nu$ decays and determination of $|V_{ub}|$. *Phys. Rev.* **D83**: 032007. [arXiv:1005.3288].

41. Ha, H. *et al.* (BELLE Collaboration). 2011. Measurement of the decay $B^0 \to \pi^- \ell^+ \nu$ and determination of $|V_{ub}|$. *Phys. Rev.* **D83**: 071101. [arXiv:1012.0090].

42. Laiho, J., E. Lunghi. & R. Van de Water. 2012. Flavor physics in the LHC era: the role of the lattice. *PoS LAT.* **2011**: 018. [arXiv:1204.0791].

43. Antonelli, M. *et al.* (FlaviaNet Working Group on Kaon Decays Collaboration). 2010. An evaluation of $|V_{us}|$ and precise tests of the standard model from world data on leptonic and semileptonic kaon decays. *Eur. Phys. J.* **C69**: 399–424. [arXiv:1005.2323].

44. Dawson, C. 2006. Progress in kaon phenomenology from lattice QCD. *PoS LAT.* **2005**: 007.

45. Antonio, D. *et al.* (RBC and UKQCD Collaboration). 2008. Neutral kaon mixing from 2 + 1 flavor domain wall QCD. *Phys. Rev. Lett.* **100**: 032001. [hep-ph/0702042].

46. Aubin, C., J. Laiho, & R. S. Van de Water. 2010. The Neutral kaon mixing parameter $B_K \to B_K$ from unquenched mixed-action lattice QCD. *Phys. Rev.* **D81**: 014507. [arXiv:0905.3947].

47. Antonelli, M. *et al.* 2010. Flavor physics in the Quark sector. *Phys. Rept.* **494**: 197–414. [arXiv:0907.5386].

48. Buras, A.J. & D. Guadagnoli. 2009. On the consistency between the observed amount of CP violation in the K^- and B_d-systems within minimal flavor violation. *Phys. Rev.* **D79**: 053010. [arXiv:0901.2056].

49. Lenz, A. *et al.* (CKMtter Collaboration). 2011. Anatomy of new physics in B-$\bar{B}$ mixing. *Phys. Rev.* **D83**: 036004. [arXiv:1008.1593].

50. Lunghi, E. & A. Soni. 2011. Possible evidence for the breakdown of the CKM-paradigm of CP-violation. *Phys. Lett.* **B697**: 323–328. [arXiv:1010.6069].

51. Bona, M. *et al.* (SuperB Collaboration). SuperB: A High-Luminosity Asymmetric e+ e- Super Flavor Factory. Conceptual Design Report. arXiv:0709.0451.

52. Isidori, G., Y. Nir. & G. Perez 2010. Flavor physics constraints for physics beyond the standard model. *Ann. Rev. Nucl. Part. Sci.* **60**: 355. [arXiv:1002.0900].

53. Brod, J., M. Gorbahn, & E. Stamou. 2011. Two-loop electroweak corrections for the $K \to \pi \nu \nu$ Decays. *Phys. Rev.* **D83**: 034030. [arXiv:1009.0947].

54. Soni, A., A.K. Alok, A. Giri, *et al.* 2010. SM with four generations: selected implications for rare B and K decays. *Phys. Rev.* **D82**: 033009. [arXiv:1002.0595].

55. Buras, A.J. 2010. Minimal flavour violation and beyond: towards a flavour code for short distance dynamics. *Acta Phys. Polon.* **B41**: 2487–2561. [arXiv:1012.1447].

56. Straub, D.M. New physics correlations in rare decays. arXiv:1012.3893.

57. Anelli, G. *et al.* (NA62 Collaboration). 2005. Proposal to measure the rare decay $K^+ \to \pi^+ \nu \bar{\nu}$ at the CERN SPS. Available at: http://na62.web.cern.ch/na62/Documents/Proposal%20spsc-2005-013.pdf.

58. Comfort, J. *et al.* (KOTO Collaboration). 2006. Proposal for $K_L \to \pi^0 \nu \bar{\nu}$ Experiment at J-Parc. Available at: http://osksn2.hep.sci.osaka-u.ac.jp/%7Etaku/jparcKL/jparc_E14 _proposal.pdf.

59. Comfort, J. *et al.* (ORKA Collaboration). 2011. ORKA: measurement of the $K^+ \to \pi^+ \nu \bar{\nu}$ Decay at Fermilab. Available

at: http://projects-docdb. fnal.gov/cgi-bin/RetrieveFile? do-
cid=1365; filename=kpnn_proposal_full.pdf;version=4.

60. Qiu, S.-W. *et al.* (Fermilab Lattice and MILC Collaboration).
2011. Semileptonic *B* to *D* decays at nonzero recoil with 2 +
1 flavors of improved staggered quarks. *PoS LAT.* **2011:** 289,
[arXiv:1111.0677].

61. Stockinger, D. 2007. The Muon magnetic moment and supersymmetry. *J. Phys. G* **G34:** R45, R92. [hep-ph/0609168].

62. Davoudiasl, H., J. Hewett, & T. Rizzo. 2000. The (*g* − 2)
of the muon in localized gravity models. *Phys. Lett.* **B493:**
135–141. [hep-ph/0006097].

63. Czarnecki, A. & W. J. Marciano 2001. The Muon anomalous
magnetic moment: a Harbinger for [new physics]. *Phys. Rev.*
D64: 013014. [hep-ph/0102122].

64. Lafaye, R., T. Plehn, M. Rauch & D. Zerwas. 2008.
Measuring supersymmetry. *Eur. Phys. J.* **C54:** 617–644.
[arXiv:0709.3985].

65. Allanach, B. *et al.* 2002. The Snowmass points and slopes:
benchmarks for SUSY searches. *Eur. Phys. J.* **C25:** 113–123.
[hep-ph/0202233].

66. Appelquist, T. & B.A. Dobrescu. 2001. Universal extra dimensions and the muon magnetic moment. *Phys. Lett.*
B516: 85–91. [hep-ph/0106140].

67. Bennett, G. *et al.* (Muon g − 2 Collaboration). 2006. Final report of the Muon E821 anomalous magnetic moment measurement at BNL. *Phys. Rev.* **D73:** 072003. [hep-
ex/0602035].

68. Davier, M., A. Hoecker, B. Malaescu. & Z. Zhang. 2011.
Reevaluation of the Hadronic contributions to the Muon *g*
− 2 and to α(*M*$_Z$). *Eur. Phys. J.* **C71:** 1515, [arXiv:1010.4180].

69. Kinoshita, T. & M. Nio. 2006. Improved α^4 term of the electron anomalous magnetic moment. *Phys. Rev.* **D73:** 013003.
[hep-ph/0507249].

70. Czarnecki, A., W. J. Marciano & A. Vainshtein. 2003. Rene-
ments in electroweak contributions to the muon anomalous magnetic moment. *Phys. Rev.* **D67:** 073006. [hep-
ph/0212229].

71. Prades, J., E. de Rafael & A. Vainshtein. Hadronic light-
by-light scattering contribution to the Muon anomalous
magnetic moment. arXiv:0901.0306.

72. Carey, R. *et al.* (The New (g − 2) Collaboration). 2010.
The new (g − 2) experiment: a proposal to measure
the Muon anomalous magnetic moment to 0.14 ppm
precision. Available at: http://gm2.fnal.gov/public_docs/
proposals/Proposal-APR5-Final.pdf.

73. Feng, X., K. Jansen, M. Petschlies. & D. B. Renner. 2011.
Two-flavor QCD correction to lepton magnetic moments
at leading-order in the electromagnetic coupling. *Phys. Rev.
Lett.* **107:** 081802, [arXiv:1103.4818].

74. Aubin, C. & T. Blum. 2007. Calculating the hadronic vacuum polarization and leading hadronic contribution to the
muon anomalous magnetic moment with improved staggered quarks. *Phys. Rev.* **D75:** 114502. [hep-lat/0608011].

75. Boyle, P., L. Del Debbio, E. Kerrane & J. Zanotti. Lattice
determination of the Hadronic contribution to the Muon *g* −
2 using dynamical domain wall fermions. arXiv:1107.1497.

76. Cohen, S. D., H.-W. Lin, J. Dudek & R. G. Edwards. 2008.
Light-meson two-photon decays in full QCD. *PoS LAT.*
2008: 159. [arXiv:0810.5550].

77. Shintani, E., S. Aoki, S. Hashimoto, *et al.* (JLQCD Collaboration). 2009. 0 to two-photon decay in lattice QCD. *PoS
LAT.* **2009:** 246. [arXiv:0912.0253].

78. Hayakawa, M., T. Blum, T. Izubuchi & N. Yamada. 2006.
Hadronic light-by-light scattering contribution to the muon
g − 2 from lattice QCD: methodology. *PoS LAT.* **2005:** 353.
[hep-lat/0509016].

Ann. N.Y. Acad. Sci. ISSN 0077-8923

Quo vadis, data privacy?

Johannes Gehrke

Cornell University, Ithaca, New York

Address for correspondence: Johannes Gehrke, Cornell University, Ithaca, NY 14853. johannes@cs.cornell.edu

Society can gain much value from Big Data. We can study census data to learn where to allocate public resources, medical records from hospitals to fight diseases, or data about students and teachers to evaluate the effectiveness of various approaches to learning and teaching. In all of these scenarios, we need to limit statistical disclosure: we want to release accurate statistics about the data while preserving the privacy of the individuals who contributed it. This paper gives an overview of recent advances and open challenges in the field, focusing on methods that probably limit how much an adversary can learn from a data release.

Keywords: data privacy; randomization; disclosure control

For reasons of modesty, fear of being thought of bigoted, or merely a reluctance to confide secrets to strangers, many individuals attempt to evade certain questions put to them by interviewers.
—Stanley Warner[57]

Introduction

The Commodore 64, the best-selling personal computer of all time, just celebrated its 30th birthday. It had 64 KB of RAM, and you could buy an external disk drive that used floppy disks with a capacity of a whopping 170 KB. At the same time the typical business computer, the IBM XT, had a hard drive with a capacity of 10 MB. Today, the standard size of a hard disk for a desktop computer is 1 TB, a factor 100,000 larger than the 10 MB. We see a similar growth in data acquisition technologies. Just as an example, the 1996 Sony Cybershot, one of the first digital cameras, had 350,000 pixels. Today's cameras have 14–18 mega-pixels, a factor of 50 larger. The associated proliferation of devices is even more staggering. About seven million new iPhones and Android devices were activated on Christmas Day in 2011, and over one billion associated applications were downloaded.

The flood of data generated by all these devices is enormous. More than 250 million pictures are uploaded to Facebook per day. The human genome has about 2.9 billion base pairs, and it took over 10 years to decode the first sequence. The DNA sequences of about 10,000 people were sequenced in 2011, and there will be millions of people in a few years. These large data sets, joined with the capabilities to perform sophisticated analysis tasks that tease nuggets out of these data, have the potential to generate tremendous value for our society; the area of Big Data has arrived.[14,49] A recent McKinsey Report estimates that there are about $300 billion in annual value to the U.S. health care system in Big Data,[52] and the discovery of Vioxx' adverse drug effects through the analysis of a large patient database by Kaiser Permanente are often cited as prime examples. The same report predicts that by 2020, personal location data will result in as much as $700 billion in value to its users through smart routing, automotive telematics, insurance pricing, and urban planning, just to name a few.

While an individual datum can be important in isolation, large data sets allow researchers to study properties of the population. The DNA of a single person has value, but once we can pool the DNAs of many people, we can study the impact of environmental factors on phenotypes and genetic variations. Similarly, a single census record by itself is of little interest, but statistics mined about all inhabitants of New York State can tell us much about shifts in socioeconomics and population, and it can

doi: 10.1111/j.1749-6632.2012.06630.x

guide policy makers toward making the right future investments. However, while there is tremendous value in Big Data, one of the challenges is that in many scenarios these data contain sensitive information; the flip side of sharing data is the potential of disclosing sensitive information.

In some cases, data sharing may be mandated. Government agencies worldwide are required to release statistical information about population, education, health, crime, and economic activities. In the United States, protecting these data goes back to the 19th century. Carrol Wright, the first head of the Bureau of Labor Statistics argued that protecting the confidentiality of the Bureau's data was necessary: if enterprises feared that data collected by the bureau would be shared with competitors, investigators, or the tax authorities, data quality would severely suffer.[37] This principle has survived the centuries, and statistical agencies today are very careful to share data without identifying data providers.[6,17] For example, today, it is the mission of the Census Bureau to "honor privacy [and] protect confidentiality."[56] But how are privacy and confidentiality achieved, and what does it formally and quantifiably mean to protect privacy?

This paper focuses on methods for sharing data sets while preserving the privacy of the entities who contributed the data. We consider the setting where a trusted *data collector* has accumulated a data set with (unaltered) private information about *data providers*, and where the data collector wants to share statistical information about this data set while limiting disclosure about individuals toward an *adversary*. The data collector does this by applying a publishing *algorithm* or *mechanism* to the data to achieve *confidentiality*, the property that an adversary cannot obtain sensitive information from the published data. *Disclosure* is best described by the famous words of Tore Dalenius: "If the release of the statistics S makes it possible to determine the value [of some private information] more accurately than is possible without access to S, a disclosure has taken place."[18] We describe methods for controlling disclosure.[21,29,61] We also consider the setting of an untrusted data collector, where it falls upon the data provider to limit disclosure to the data collector.

Failure to avoid disclosure can result in much damage. Text boxes 1 and 2 show the difficulty of sharing data while preserving privacy, and similar observations have been made about a lot of other

---

The AOL Search Log Release. In August 2006 America Online (AOL) wanted to do a "good thing," and it released 20 million search queries to the research community.[7,58] These queries were posed by over 650,000 users over a period of three months. AOL had replaced the user names with random numbers, but many of the queries included personal information about their users (such as location queries, or simply vanity searches).[59] Only three days after the release, reporters from the New York Times linked one of the supposedly anonymized users in the data to Mrs. Thelma Arnold, a lady living in Lilburn, Georgia.[8] This example shows that armed with a small amount of public information — in the case of the reporters from the New York Times simply a phone book — we can link the supposedly anonymized IDs with real people, thus revealing their history of searches, their part of the "database of intentions".[9] Even stronger forms of anonymization do not provide stronger protection if the adversary has sufficient background information. For example, assume we would also anonymize every keyword in the search log by using a one-way hash function to map it to a random number. It has been shown that even in this case an adversary, who has access to another search log, can invert the one-way function through frequency attacks.[46] It is possible to publish search logs with very strong privacy guarantees, but only by using very different methods.[38,45]

Text box 1. Privacy breach, example 1: AOL search log release.

---

types of data.[40,51,63] Sharing data has the promise of creating huge value, but how can we share data without running into any of these disaster scenarios? It is the goal of this paper to give a brief introduction to this field, also called *statistical disclosure control*[29] or *privacy-preserving data publishing*.[16,35]

The Netflix Prize. In 2006, Netflix released a dataset containing 100 million anonymized movie ratings of its customers as part of "The Netflix Prize."[13] The release was a challenge to the research community: Use this data to design a movie recommendation algorithm that would beat Netflix' existing recommendation algorithm by at least 10%. Due to the associated grand prize of $1,000,000 promised to the winner, over 5,000 teams from all over the world competed over the following three years to win the bounty.[a] The data was anonymized by replacing customer names with random numbers and by perturbing some of the rating data. However, these methods turned out to be insufficient: Just a year later, researchers linked some of the Netflix data with movie ratings from the Internet Movie Database (IMDB) through a comparison of ratings and their timestamps. Suddenly, other ratings, that these users did not choose to make public on IMDB, were out in the open.[34,54]

[a]The race for the Netflix Prize is a fascinating story by itself .[10–12, 60]

Text box 2. Privacy breach, example 2: the Netflix prize.

The past

Randomized response

Let us jump backward to 1960, when Stanley Warner was faced with a privacy conundrum. During interviews for market surveys, an individual, say Alice, who is the data provider in this scenario, is asked by the data collector called Bob a series of yes/no questions. But Alice may refuse or not to honestly answer questions of a sensitive or controversial nature—she does not trust Bob and is worried about being discriminated against because of what the true answers would be.[57] Warner's idea, called *randomized response*, was to let Alice flip a biased coin without showing the outcome of the coin flip to Bob. If the coin comes up heads, Alice answers the question directly; if the coin comes up tails, she responds with the negation of her answer. Since Bob does not know the outcome of the coin flip, Bob does not know whether he received truthful answers, and thus this method intuitively protects Alice's privacy; her answer could always have been due to the coin flipping on the other side.

Let us make this argument a bit more formal. Consider a simple "yes" or "no" question Q. Let π be the true probability that a randomly selected person from the population answers "yes" to question Q, and let p be the probability that the coin in Warner's method comes up "heads." Now assume we are given a sample with n such randomized answers where m respondents have answered "yes." Then the maximum likelihood estimate for π is

$$\hat{\pi} = \frac{(p-1)}{(2p-1)} + \frac{m}{n(2p-1)}.$$

It is easy to show that this is an unbiased estimator with $E(\hat{\pi}) = \pi$ and that its variance has two components, the regular sampling variance and an additional component due to coin flips:

$$Var(\hat{\pi}) = \frac{\pi(1-\pi)}{n} + \frac{\frac{1}{16(p-0.5)^2} - \frac{1}{4}}{n}$$

$$= \text{sampling variance} + \text{coin flips}.$$

Thus, unless the coin is unbiased ($p = 0.5$), we can compute an unbiased estimator, $\hat{\pi}$ for π. Also, the variance of $\hat{\pi}$ has not grown out of bounds; it contains the regular sampling variance, but also an additional component that increases the variance due to the randomness in the coin flips.

In practice this means that we need to interview more persons in order to estimate π as precisely as without randomized response. But how much privacy is really preserved by randomized response? Against what types of adversaries are we defended?

Over the next decades, researchers developed various heuristics with the goal to limit disclosure. Statisticians defined different metrics of risk, such as small counts in contingency tables,[28] using identification rules[47] or decision-theoretic approaches,[23] and various methods for limiting disclosure, such as many sophisticated variants of randomized response,[57] sampling,[30,31] suppression and swapping,[32] generalization or top-coding, synthetic data generation,[3,4,15,22] data perturbation, and the publishing of marginals in contingency tables, just to name a few.[16] Methods such as data swapping, top-coding, and subsampling are the methods of choice of statistical offices today.[19] Swapping means that we take two records in the database and swap the values of one or more attributes; this technique was used in the United States for the 1990 and 2000 censuses.[33] In top-coding, we change the values of some outliers (in practice, often at the bottom or top end of the attribute values) to some more general value. In subsampling, we do not publish every record, but rather sample records with a (possibly nonuniform) sampling rate. Many of these methods have been implemented in software packages.[41,42] In addition to changing the data that is published, in practice, statistical offices limit disclosure further by keeping the parameters of the algorithms, such as the actual swap rate or the sampling rate secret.

However, while many ingenious methods for limiting disclosure in data publishing were developed,[31,39,48] a formal connection between the power of an adversary and methods that limit statistical disclosure remained open.

The present

Fast forward 40 years since Stanley Warner. Rakesh Agrawal, in his innovations award talk at the 1999 Annual ACM SIGKDD International Conference, phrased the question of limiting disclosure as the challenge of privacy-preserving data mining: How can we mine a data set without leaking any personal data about the people in the data set?[5] The computer scientists who were intrigued by this question especially wanted formal models of data privacy—mathematical definitions of information leakage and adversary models like those in computer security, especially cryptography.

Privacy breaches

One of the first problems addressed by the computer science community was a variant of Stanley Warner's problem in the context of recommender systems. Assume that Bob would like to expand his activities and build a service that shows what books people frequently purchase together—similar to Amazon's "people who purchased this item also purchased that item." Assume that we have a universe of say 100,000 books. Our problem can be phrased as Alice sending a list of zeros and ones to Bob, where a zero at position i indicates that she did not purchase Book i, and a one indicates a purchase. However, assume that in order to preserve her privacy, Alice does not want to simply send her whole purchase history to Bob. So, similar to Warner's scenario, maybe Alice can send a randomized version of her history to the service?

Let us apply a straightforward extension of Warner's technique to this problem: Instead of Alice sending her unaltered list, she randomizes it by taking each entry that contains a one and with probability $P = 80\%$, she sets it to zero and instead sets another random entry to one. How much privacy does this algorithm give us? Since there is a high likelihood (80%) that a one-bit will be flipped to a zero-bit, we may think that this randomization preserves privacy and that Bob who sees Alice's randomized list cannot make any strong conclusions about Alice's original list.

However, this method (maybe surprisingly) does not preserve the Alice's privacy. Consider a set of say four unique books that only Alice purchased together. The probability that any other list of books after randomization contains exactly these four books is extremely small, since there are 100,000 books in total. Let us illustrate this phenomenon with an example. Assume that there are 10 million people who have each purchased 10 books from our universe of 100,000 books. Consider a specific combination of three books, "*Harry Potter and the Philosopher's Stone*," "*Harry Potter and the Chamber of Secrets*," and "*Harry Potter and the Prisoner of Azkaban*" (let us abbreviate them H1, H2, and H3, respectively) and assume that 1% of customers

purchased all three of them (we will abbreviate this H123), 5% of customers purchased any two but not all three, and 94% of customers purchased only one or zero of these three books.

Let us now apply the above variant of Warner's technique. After randomization, there will be, in expectation, 814 lists that contain H123. Out of these 814 lists, 800—or about 98%—contained H123 originally! This means, that even though the prior probability of Alice having purchased H123—the probability of the list of a persons containing H123 without having seen the randomized list of a person—is only 1%, the posterior probability, after having seen that the randomized list of Alice contains H123, jumps to over 98%! The intuition is as follows: the probability that the triple H123 is generated through randomization is very low; thus, whenever we see H123, even in a randomized transaction, it is very unlikely that it was generated by chance. We call this jump between prior and posterior probability a *privacy breach*.[27] Through the data release, Bob can sometimes associate some properties to Alice with a much higher probability as compared to without the data release.

How can we avoid privacy breaches? Consider a randomization operator $\mathcal{A}$ that Alice uses to randomize her data before sending it to Bob. We can now formulate a strong privacy condition for such an algorithm:

DEFINITION 1 (AT MOST γ-AMPLIFYING PRIVACY).[15] *A randomized algorithm $\mathcal{A}$ satisfies γ-amplifying privacy, if for all input records r_1, r_2, and for any output record s,*

$$\frac{P(\mathcal{A}(r_1) = s)}{P(\mathcal{A}(r_2) = s)} \leq \gamma,$$

where the probability is over the randomness of $\mathcal{A}$.

It can be shown that if Alice applies an algorithm $\mathcal{A}$ that is at most γ-amplifying, then privacy breaches will never occur. Assume that Bob wants to know a property T about Alice, and that Bob has a prior belief about $P(T)$. Then we can show that Bob's posterior belief $P(T|\mathcal{A}(r))$ about r, given $\mathcal{A}(r)$, is bounded by his prior belief as follows:

$$\frac{P(T|\mathcal{A}(r))}{1 - P(T|\mathcal{A}(r))} \leq \gamma \frac{P(T)}{1 - P(T)}.$$

For example, assume that $\frac{P(T)}{1-P(T)} = 10$; this means that Bob, prior to seeing Alice's response, believes that it is 10 times more likely that Alice has the property T than not. After seeing Alice's randomized record $\mathcal{A}(r)$, his belief has grown, at most, by a factor of γ. In other words, if Alice applies an algorithm that is at most γ-amplifying and then shares her modified record with Bob, Bob is not going to learn much more about Alice than he already knew. The parameter γ is selected such that Alice has sufficient privacy but Bob can gain some knowledge. This prevents Bob from learning much about individual data providers, such as Alice, but he can learn properties about the population.

Differential privacy

Much attention has also been focused on the *data publishing* case where the data collector is trusted and thus obtains true, unaltered records from the data providers. The data collector then publishes these data while limiting statistical disclosure; for example, the United States Census Bureau directly matches this setup. Out of the early work in this setting, a beautiful and clean privacy definition, called *differential privacy*, emerged, and we will motivate and introduce it in this section.

How would we extend privacy breaches to the data publishing scenario? A straightforward extension of Definition 1 from individual records to whole data sets would be to replace the notion of input records with data sets, resulting in the following definition, called *free-lunch privacy*.[43]

DEFINITION 2 (FREE-LUNCH PRIVACY).[43] *A randomized algorithm $\mathcal{A}$ provides ϵ-free-lunch privacy, if for any pair of databases D_1, D_2 and for any set S,*

$$P(\mathcal{A}(D_1) \in S) \leq e^\epsilon P(\mathcal{A}(D_2) \in S),$$

where the probability is over the randomness of $\mathcal{A}$.

Given this strong definition, it is easy to see that an algorithm that satisfies free-lunch privacy also ensures that no privacy breaches occur. Unfortunately, however, free-lunch privacy is way too strong, as it does not provide much utility in the resulting data set. Let us give one very simple notion of utility that illustrates this point well.

DEFINITION 3 (DISCRIMINANT-UTILITY).[43] *Given an integer $k > 1$, a randomized algorithm $\mathcal{A}$, and a constant c, the discriminant $\omega(k, \mathcal{A})$ is the largest constant c, such that there exist k possible databases $D_1, \ldots, D_k$ and disjoint sets $S_1, \ldots, S_k$, such that $P(\mathcal{A}(D_i) \in S_i) \geq c$ for $i = 1, \ldots, k$.*

The intuition is that for an $\mathcal{A}$ and k, if the discriminant $\omega(\mathcal{A}, k)$ is close to $1/k$, then we cannot get any useful answers from the output of $\mathcal{A}$, since the output of $\mathcal{A}$ is uniformly distributed independently of its input database. It can be easily shown that the discriminant of any algorithm that satisfies free lunch privacy is at most $\frac{e^\epsilon}{k-1+e^\epsilon}$, which is basically $1/k$. Thus, with free-lunch privacy, we cannot distinguish much between *any* two data sets, whether they are small or large, or whether they contain only healthy people or not. Thus instead of relating any two data sets as in free-lunch privacy, differential privacy restricts Definition 2 to "adjacent" data sets that differ in only one record.

DEFINITION 4 (DIFFERENTIAL PRIVACY).[26] *A randomized algorithm A provides ϵ-differential privacy if, for all data sets D and D' that only differ in one record, and all $S \subseteq Range(\mathcal{A})$, we have that*

$$P[\mathcal{A}(D) \in S] \leq e^\epsilon \cdot P[\mathcal{A}(D') \in S],$$

where the probability is over the randomness of $\mathcal{A}$.

Note that differential privacy does not hide any details about the algorithm $\mathcal{A}$, nor does it hide its parameters; the only hidden information is the random choices made by the algorithm.

However, differential privacy needs another important assumption, which we illustrate through the adaptation of an example from our recent work.[36] Consider a social network of people that are grouped into cliques of size 200. In each clique, either at least 80% of the people are members of the New York Academy of Sciences (NYAS) or at least 80% are not members. Overall, assume that the importance of science has been realized widely enough that the number of members is about the same as the number of nonmembers in the population. Now if we were to use a mechanism that would provide differential privacy when publishing the value of the membership attribute, the mechanism would add enough noise to protect against deducing the NYAS membership information of any single person. However, once we release the data, we can also see the connections in the social network, and thus membership in a clique, which allows us to deduce whether a person is a member, with probability close to 80%.

This example shows that differential privacy requires independence among records in the database: even though differential privacy hides the membership of Alice, we can conclude her membership status from the membership of Alice's friends! In general, it has been proven that if we make no assumptions about the mechanism P that generates the data set D that we want to protect, then for any data publishing algorithm $\mathcal{A}$ that achieves privacy and some utility, there exists a P that blatantly breaches privacy! In other words, unless we make some assumptions about P, we cannot publish any data from P.[24,43] Thus we need to make some assumptions about P.

It can be shown that differential privacy corresponds to restricting P to distributions over independent tuples, and thus it can achieve some utility. Note that this does not usually hold in practice, as there are often correlations among some records in the database. We will elaborate on this issue in "Differential privacy is too weak."

Since the original proposal of differential privacy, much progress has been made in the development of mechanisms that achieve differential privacy;[25] for example, if we are only interested in simple histograms about the data, we can achieve differential privacy by adding a small amount of noise sampled from an appropriately scaled distribution, such as the Laplace or Gaussian distribution. The national statistical offices have also started to pay attention, and there now exist public data products that are published with formal privacy guarantees, as we will discuss next.

A success story: OnTheMap

We discuss an application based on the U.S. Census Bureau's Longitudinal Employer-Household Dynamics (LEHD) Program;[1,2,55] our discussion is an abbreviation of the full description of the method by Machanavajjhala *et al.*[50] The application, called OnTheMap, shows where workers live and are employed; it contains a table "Commute_Patterns" with schema *(id, origin_block, destination_block)*, where each row represents a worker. The attribute *id* is a random number serving as a key for the table, *origin_block* is the census block in which the worker lives, and *destination_block* is where the worker works. An origin block o corresponds to a destination block d, if there is a tuple with *origin_block* o and *destination_block* d. For example, with these data, users can visualize for a destination block d points on a map that show the corresponding origin blocks from which workers commute to d. Note that there are about 10 million census blocks,

1. We have an input dataset of origin-block, destination-block pairs; our method of generating the output dataset for publication is *synthetic data generation*.

2. We learn a statistical model from the input data. We build the model such that it will overfit the data since we want to capture as many characteristics of the input data as possible. We generate the output data by sampling from the model. Privacy is achieved during the creation of the model and due to random sampling from the model.

3. We want to use a formal privacy definition. If we were to use differential privacy, we would have to defend ourselves against very unlikely worst-case privacy breaches and thus add a huge amount of noise. We thus resort to probabilistic differential privacy, where differential privacy is achieved except for a very small probability that the publication results in a disclosure.

4. Despite our relaxation of differential privacy, we still run into two issues with the utility of the published data. First, it is possible (albeit with very small probability) that in the synthetic data, all the workers commuting to New York City come from Boston (or worse, from San Francisco) even though this is not the case in the original data. Thus we use accept/reject sampling trying until the (randomized) algorithm generates a dataset that is representative of the input. Second, due to the large number of census blocks, the total amount of noise required to guarantee privacy may swamp most of the signal. For this we had to develop a special method that hierarchically shrinks the number of origin blocks by grouping blocks that are far away from the origin block and thus reducing the total amount of noise added.

Text box 3. Steps in synthetic data generation for OnTheMap.

so the domain is very large and the data are very sparse.

Information about destination blocks had already been publicly released, thus only the *origin_block* is sensitive, and an adversary should not be able to identify individuals with specific origin blocks. The details of the full algorithm are beyond the scope of this paper, but Text box 3 outlines the steps that we had to go through in order to limit disclosure while preserving some utility.[50] Since version 3, OnTheMap has now been published with a variant of this method. The lessons from this discussion are threefold. First, formal privacy definitions have been successful, in that OnTheMap is now protected against very powerful adversaries that know everybody except one person in the data set. Second, unfortunately there was no out-of-the-box method that we could apply; it required the development and careful analysis of several new steps, such as the accept/reject sampling and the coarsening of the domain. Third, it is not clear that we need to defend against such a powerful adversary as assumed in differential privacy; if we had a formal privacy definition that would only defend against a weaker adversary (who, e.g., would only know up to a constant number of people), it would probably be sufficient for this application.

The future

As we discussed in "The present section," differential privacy is currently the state-of-the-art in terms of applying formal privacy guarantees to real problems. However, although differential privacy provides a strong privacy guarantee, there are settings where its underlying assumption of independence between records in the database does not hold. In addition, there are scenarios where differential privacy is clearly too strong, and thus we need definitions that only guard us against weaker adversaries than differential privacy. We discuss each of these scenarios now in turn, and then we turn to the larger question of building useful tools for nonexperts.

Differential privacy can be too weak

Recall our example from "Differential Privacy" section, where we tried to hide information about membership about the NYAS, but since membership was not independent of the membership of a person's friends, differential privacy could not hide this information. This was due to the fact that a core assumption in differential privacy does not hold: the random process that clusters people into cliques of members and nonmembers violates the assumption of differential privacy that requires that people are independent. Thus since the structure of the social network provides auxiliary information about each person, a person's privacy can be violated even under differential privacy!

How do we overcome this restriction? One recent approach relies on the notion of zero-knowledge from cryptography, which basically states that an adversary who obtains an answer from a mechanism (such as an encrypted string) gains zero additional knowledge from the encrypted string; informally, the adversary is only able to compute as much as without having access to the encrypted string. Note that such a strong notion of privacy would be useless in practice: we gain no utility

from publishing the data set, since we can only learn as much as without publishing it. Thus we need to give the mechanism some power in order to release some useful data. We use the notion of a *simulator* from zero-knowledge; a simulator is an algorithm that can compute something useful given some information. We now can (informally) say that a mechanism achieves zero-knowledge privacy if it can compute no more than a simulator that can access only aggregate information about the data set.

With this definition, we can now characterize the functions that can be computed from a release with zero-knowledge privacy; in particular, we characterize them in terms of their sample complexity, that is, how well such functions can be approximated just using random samples from the data set. Many interesting functions that can be computed have low sample complexity, including averages, sums, and coarse histogram queries; these are the types of statistics that zero-knowledge privacy releases with good utility. In addition, we can also release information about the structure of social networks with zero-knowledge privacy, such as the average degree and properties such as connectivity and cycle-freeness.[36]

Differential privacy can be too strong

As we have seen in the previous section, there are scenarios where differential privacy is too weak, but there are also scenarios where differential privacy is too strong. In most real scenarios assuming that an adversary knows all but one record in the database is highly unrealistic; thus adding noise sufficient to achieve differential privacy destroys too much utility in the data. For example, both survey and census data often have more than 100 attributes, many of them binary. Ensuring differential privacy in a 100-dimensional space will drown all signal in the data unless the data are extremely well clustered.

In addition, social networks are only one example of where the assumptions of differential privacy do not hold; other examples include any types of correlations between data providers (such as belonging to the same family) or prior publication of statistics about the data set that no longer leave records independent.[44] So, in general, there are data (e.g., as in the previous example graph data) where parts (such as the edges in a graph) are considered private, but where applications of differentially private mechanisms directly to the data would render the resulting structure useless. For example, we currently can publish the degree distribution of a social network with differential privacy, but not the graph itself. How to publish such rich structures, such as graphs with formal privacy guarantees, is an open problem.

Beyond new privacy definitions

This paper mainly discussed the ongoing search for the right privacy definition. Although differential privacy or a variant is a prime candidate, the jury is still out. This lack of a consensus also reflects the state-of-the-art in existing tools. Developing OnTheMap, as described in "A success story: OnTheMap" took a group of scientists over a year; we could not apply existing techniques, and we had to invent novel methods to achieve our goal. There are the first interesting steps toward integrating differential privacy into the programming language LINQ,[53] and we are unfortunately still far away from a set of tools that could be usable by statistical offices or other users that are facing data publishing problems.

Conclusions

This paper gives only a glimpse into a new exciting area at the confluence of ideas from computer science, statistics, law, and the social sciences. I believe we will see much further progress on formal privacy definitions and improved methods, and I hope that future data products from the national statistics offices will be published with some formal notion of disclosure control. Carrol Wright would be amazed by the field today.

Acknowledgments

The author would like to thank Alan Demers, Michael Hay, Ashwin Machanavajjhala, and the anonymous reviewers for helpful comments. The author gratefully acknowledges generous support from the Blavatnik Award for Young Scientists from the New York Academy of Sciences. The research described in this article was supported by the National Science Foundation (NSF) under Grant IIS-1012593. Any opinions, findings, and conclusions or recommendations expressed in this material are those of the author and do not necessarily reflect the views of the sponsors.

Conflicts of interest

The author declares no conflicts of interest.

References

1. Abowd, J.M. & F. Kramarz. 1999. Econometric analysis of linked employer-employee data. *Labour Econ.* **6:** 53–74.
2. Abowd, J.M., F. Kramarz & D. Margolis. 1999. High wage workers and high wage firms. *Econometrica* **67:** 251–333.
3. Abowd, J.M. & J.I. Lane. 2004. New approaches to confidentiality protection: synthetic data, remote access and research data centers. In *Privacy in Statistical Databases: CASC Project International Workshop, PSD 2004. Proceedings, volume 3050 of Lecture Notes in Computer Science.* J. Domingo-Ferrer & V. Torra, Eds.: 282-289. Springer-Verlag, Berlin.
4. Abowd, J.M. & L. Vilhuber. 2008. How protective are synthetic data? In *Privacy in Statistical Databases.* J. Domingo-Ferrer & Y. Saygin, Eds.: 239–246. Vol. 5262 of *Lecture Notes in Computer Science.* Springer.
5. Agrawal, R. 1999. Keynote: Data mining: crossing the chasm. In *Proceedings of the 5th ACM SIGKDD International Conference on Knowledge Discovery and Data Mining (KDD-99).*
6. Anderson, M. & W. Seltzer. 2007. Challenges to the confidentiality of U.S. federal statistics, 1910-1965. *J. Off. Stat.* **23:** 1–34.
7. Arrington, M. 2006. AOL proudly releases massive amounts of private data. TechCrunch: http://www.techcrunch.com/2006/08/06/aol-proudly-releasesmassive-amounts-of-user-search-data/, August 2006.
8. Barbaro, M. & T. Zeller. 2006. A face is exposed for AOL searcher no. 4417749.
9. Battelle, J. 2005. *The Search: How Google and Its Rivals Rewrote the Rules of Business and Transformed Our Culture.* Portfolio Hardcover, US.
10. Bell, R., J. Bennett, Y. Koren & C. Volinsky. 2009. The million dollar programming prize. *IEEE Spect.* **5:** 28–33.
11. Bell, R.M. & Y. Koren. 2007. Lessons from the Netflix prize challenge. *SIGKDD Explor.* **9:** 75–79.
12. Bell, R.M., Y. Koren & C. Volinsky. 2007. Chasing $1,000,000: how we won the Netflix progress prize. *Stat. Comput. Stat. Graph. Newslett.* **18:** 4–12.
13. Bennett, J. & S. Lanning. 2006. The Netflix prize. In *Proceedings of the KDDCup'07.*
14. Brynjolfsson, E. & A. McAfee. 2011. The Big Data Boom is the Innovation Story of Our Time. The *Atlantic.* Nov. 21.
15. Caiola, G. & J.P. Reiter. 2010. Random forests for generating partially synthetic, categorical data. *Trans. Data Privacy* **3:** 27–42.
16. Chen, B.-C., D. Kifer, K. LeFevre & A. Machanavajjhala. 2009. Privacy-preserving data publishing. *Foundations Trends Databases* **2:** 1–167.
17. Cox, L.H., S. McDonald & D. Nelson. 1986. Confidentiality issues at the U.S. Bureau of the Census. *J. Offic. Stat.* **2:** 135–160.
18. Dalenius, T. 1977. Towards a methodology for statistical disclosure control. *Statistik Tidskrift* **15:** 429–444.
19. Denning, D.E. 1980. Secure statistical databases with random sample queries. *ACM Trans. Database Syst.* **5:** 291–315.

20. Domingo-Ferrer, J. & V. Torra, Eds. 2004. *Privacy in Statistical Databases: CASC Project International Workshop, PSD 2004*. Proceedings, volume 3050 of *Lecture Notes in Computer Science*. Springer, Barcelona, Spain.

21. Doyle, O., J. Lane, J. Theeuwes & L. Zayatz, Eds. 2001. *Confidentiality, Disclosure and Data Access: Theory and Practical Application for Statistical Agencies*. Elsevier. New York, NY.

22. Drechsler, J. & J.P. Reiter. 2011. An empirical evaluation of easily implemented, nonparametric methods for generating synthetic datasets. *Comput. Stat. Data Anal.* 55: 3232–3243.

23. Duncan, G.T. & D. Lambert. 1986. Disclosure-limited data dissemination. *J. Am. Stat. Assoc.* 81: 393: 10–18.

24. Dwork, C. 2008. Differential privacy: a survey of results. In *Proceedings of the TAMC*, volume 4978 of *Lecture Notes in Computer Science*. M. Agrawal, D.-Z. Du, Z. Duan & A. Li, Eds.: 1–19. Springer, Berlin.

25. Dwork, C. 2011. A firm foundation for private data analysis. *Commun.* 54: 86–95. ACM, New York.

26. Dwork, C., F. McSherry, K. Nissim & A. Smith. 2006. Calibrating noise to sensitivity in private data analysis. In *Proceedings of the TCC*, volume 3876 of *Lecture Notes in Computer Science*. S. Halevi & T. Rabin, Eds.: 265–284. Springer, Berlin.

27. Evfimievski, A.V., J. Gehrke & R. Srikant. 2003. Limiting privacy breaches in privacy preserving data mining. In *Proceedings of the PODS*, 211–222. ACM, New York.

28. Fellegi, I.P. 1972. On the question of statistical confidentiality. *J. Am. Stat. Assoc.* 67: 337:7–18.

29. Fienberg, S.E. & J. Jin. 2009. Statistical disclosure limitation for data access. In *Encyclopedia of Database Systems*. L. Liu & M.T. Özsu, Eds.: 2783–2789. Springer, US.

30. Fienberg, S.E. & U.E. Makov. 1998. Confidentiality, uniqueness and disclosure limitation for categorical data. *J. Offic. Stat.* 14: 185–397.

31. Fienberg, S.E., U.E. Makov & A.P. Sanil. 1997. A bayesian approach to data disclosure: optimal intruder behavior for continuous data. *J. Offic. Stat.* 13(1).

32. Fienberg, S.E., U.E. Makov & R.J. Steele. 1998. Disclosure limitation using perturbation and related methods for categorical data. *Journal of Official Statistics* 14(4).

33. Fienberg, S.E. & J. McIntyre. 2004. Data swapping: variations on a theme by Dalenius and Reiss. In *Privacy in Statistical Databases* 14–29.

34. Frankowski, D., D. Cosley, S. Sen, L.G. Terveen, & J. Riedl. 2006. You are what you say: privacy risks of public mentions. In *SIGIR*. E.N. Efthimiadis, S.T. Dumais, D. Hawking & K. Järvelin, Eds.: 565–572. ACM.

35. Fung, B.C.M., K. Wang, R. Chen & P.S. Yu. 2010. Privacy-preserving data publishing: a survey of recent developments. *ACM Comput. Surv.* 42: 1–53.

36. Gehrke, J., E. Lui & R. Pass. 2011. Towards privacy for social networks: A zero-knowledge based definition of privacy. In Y. Ishai, Ed.: *TCC*, volume 6597 of *Lecture Notes in Computer Science*: 432–449. Springer.

37. Goldberg, J.P. & W.T. Moye. 1985. *The First Hundred Years of the Bureau of Labor Statistics*. Bureau of Labor Statistics, Bulletin 2235. Superintendent of Documents, U.S. Government Printing Office, Washington, DC 20402.

38. Götz, M., A. Machanavajjhala, G. Wang, X. Xiao & J. Gehrke. 2012. Publishing search logs - a comparative study of privacy guarantees. *IEEE Trans. Knowl. Data Eng.* 24: 520–532.

39. Herzog, T.N., F.J. Scheuren & W.E. Winkler. 2007. *Data Quality and Record Linkage Techniques*. Springer-Verlag, Berlin.

40. Homer, N., S. Szelinger, M. Redman, *et al.* 2008. Resolving individuals contributing trace amounts of DNA to highly complex mixtures using high-density snp genotyping microarrays. *PLoS Genet* 4: e1000167.

41. Hundepool, A. 2004. The argus software in the casc-project. In *Privacy in Statistical Databases: CASC Project International Workshop, PSD 2004. Proceedings, volume 3050 of Lecture Notes in Computer Science*. J. Domingo-Ferrer & V. Torra, Eds.: 323–335. Springer-Verlag, Berlin.

42. Hundepool, A. 2006. The argus software in cenex. In *Privacy in Statistical Databases*, volume 4302 of *Lecture Notes in Computer Science*. J. Domingo-Ferrer & L. Franconi, Eds.: 334–346. Springer-Verlag, Berlin.

43. Kifer, D. & A. Machanavajjhala. 2011. No free lunch in data privacy. In *Proceedings of the SIGMOD Conference*. T.K. Sellis, R.J. Miller, A. Kementsietsidis & Y. Velegrakis, Eds.: 193–204. ACM, New York.

44. Kifer, D. & A. Machanavajjhala. 2012. A rigorous and customizable framework for privacy. In *Proceedings of the PODS Conference*.

45. Korolova, A., K. Kenthapadi, N. Mishra & A. Ntoulas. 2009. Releasing search queries and clicks privately. In *WWW*. J. Quemada, G. León, Y. S. Maarek & W. Nejdl, Eds.: 171–180. ACM, New York.

46. Kumar, R., J. Novak, B. Pang & A. Tomkins. 2007. On anonymizing query logs via token-based hashing. Williamson *et al.* 62: 629–638.

47. Lambert, D. 1993. Measures of disclosure risk and harm. *J. Offic. Stat.* 9: 313–331.

48. Lambert, D. 1993. Measures of disclosure risk and harm. *J. Offic. Stat.* 9: 313–331.

49. Lohr, S. 2012. The age of Big Data. *New York Times*.

50. Machanavajjhala, A., D. Kifer, J. Abowd, J. Gehrke & L. Vilhuber. 2008. Privacy: from theory to practice on the map. https://courses.cit.cornell.edu/jma7/ICDE08_conference_0768.pdf

51. Malin, B.A. 2005. Technical evaluation: an evaluation of the current state of genomic data privacy protection technology and a roadmap for the future. *JAMIA* 12: 28–34.

52. Manyika, J., M. Chui, B. Brown, *et al.* 2011. *Big Data: The Next Frontier for Innovation, Competition, and Productivity*. http://www.mckinsey.com/mgi/publications/big_data/pdfs/MGI_big_data_full_report.pdf

53. McSherry, F.D. 2010. Privacy integrated queries: an extensible platform for privacy-preserving data analysis. *Commun.* 53: 9: 89–97. ACM, New York.

54. Narayanan, A. & V. Shmatikov. 2008. Robust de anonymization of large sparse datasets. In *IEEE Symposium on Security and Privacy*: 111–125. IEEE Computer Society.

55. United States Census Bureau. On the map (version 3): Longitudinal employer-household dynamics. Available at: http://onthemap.ces.census.gov/. Last accessed March 1, 2012.

56. United States Census Bureau. 2012. Mission statement. Available at: http://www.census.gov/aboutus/mission.html. Last accessed March 1, 2012.

57. Warner, S.L. 1965. Randomized response: a survey technique for eliminating evasive answer bias. *J. Am. Stat. Assoc.* **60:** 63–69.

58. Wikipedia. AOL search data leak. Available at: http://en.wikipedia.org/wiki/AOL_search_data_leak. Last accessed March 1, 2012.

59. Wikipedia. Egosurfing. Available at: http://en.wikipedia.org/wiki/Egosurfing. Last accessed March 1, 2012.

60. Wikipedia. Netflix prize. en.wikipedia.org/wiki/Netflix_Prize.

61. Willenborg, L. & T. de Waal. 2001. *Elements of Statistical Disclosure Control,* Vol. 155 of *Lecture Notes in Statistics.* Springer-Verlag, Berlin.

62. Williamson, C.L., M.E. Zurko, P.F. Patel-Schneider & P.J. Shenoy, Eds. 2007. In *Proceedings of the 16th International Conference on World Wide Web, WWW 2007,* ACM, Banff, Alberta, Canada.

63. Yeniterzi, R., J.S. Aberdeen, S. Bayer, *et al.* 2010. Effects of personal identifier resynthesis on clinical text de identification. *JAMIA* **17:** 159–168.

Ann. N.Y. Acad. Sci. ISSN 0077-8923

ANNALS OF THE NEW YORK ACADEMY OF SCIENCES
Issue: *Blavatnik Awards for Young Scientists*

Exploring the birth and death of black holes and other creatures

Szabolcs Márka[1,2]

[1]Columbia Astrophysics Laboratory, Pupin Physics Laboratories, Columbia University, New York, New York. [2]Laboratoire AstroParticule et Cosmologie (APC) Université Paris Diderot, CNRS: IN2P3, CEA: DSM/IRFU, Observatoire de Paris - 10, Paris, France

Address for correspondence: S. Márka, Columbia Astrophysics Laboratory 1009, Pupin Physics Laboratories, Mail Code 5247, 550 West 120th Street, New York, NY 10027. sm2375@columbia.edu

Astronomers and physicists of diverse interest are teaming up to study enigmatic cosmic phenomena, such as the life cycle of black holes. A "disruptive innovation" is about to emerge during the next decade: Advanced gravitational-wave observatories. The emergence of gravitational-wave physics as a viable observational channel is expected to improve our understanding of the Universe in unprecedented and plausibly unexpected ways, and to enhance the capabilities of the astrophysics community. Detecting cosmic counterparts to gravitational-wave events would revolutionize our understanding of violent astrophysical processes, such as the birth and death of black holes and neutron stars. Although the vanguard of joint observational work with electromagnetic observatories has already rewarded us with a glimpse of the power of gravitational-wave astronomy, the most interesting science is yet to come. Many sources of gravitational-waves are expected to be observable through a broad set of messengers, including γ-rays, X-rays, optical, radio, and neutrino emission. Multimessenger investigations may be crucial for the first detection of gravitational-waves, and could provide the broadest scientific impact afterwards. This paper outlines some exciting aspects of transient multimessenger astronomy with gravitational-waves and highlights open questions that might be resolvable by Advanced or third generation gravitational-wave detector networks. In addition, we will use examples from current research to illustrate that the toolkit of fundamental research can enrich other fields, and that synergistic science can expand horizons here on Earth.

Keywords: astrophysics; gravitational-waves; biophysics, multimessenger astronomy

Introduction

Gravitational-waves[1–6]—often conceptualized as the "ripples in the fabric of space-time"—are predicted by Einstein's theory of General Relativity. Although we have strong, but indirect, evidence of their existence, they had never been detected directly. In a concerted effort, over the past decades, dedicated teams of physicists have built large gravitational-wave detectors that interferometrically[7,8] monitor the minute relative displacement of test masses (mirrors) in response to local space-time distortions induced by gravitational-waves originating from extremely energetic cosmic sources. Indeed, astrophysics deduced from gravitational-wave observations can be a profoundly new source of information about these sources for the following reasons: (1) gravitational-waves interact weakly with matter, propagating through space essentially undisturbed, and thus can carry information about crucial aspects of cosmic processes and fundamental physics usually hidden from electromagnetic observations; and (2) gravitational-waves directly probe sources that go through processes with strong approximately nonspherical mass dynamics. Compact inspiraling binary systems containing neutron stars and black holes, stellar core collapses, and the dynamics of the early universe immediately following the Big Bang are primary examples[6] of energetic gravitational-wave generating processes. Among others, the community hopes to get insight on theories of alternative gravity,[9,10] on the

doi: 10.1111/j.1749-6632.2011.06414.x

hidden putative population of black holes in globular clusters and galactic cores,[11–13] and on the very existence of intermediate mass black holes.[13,14]

The quest for the first direct detection of gravitational-waves will get a boost by the middle/end of the decade as the second generation of Earth-based interferometric detector network begins collecting data, including the U.S.-based Advanced Laser Interferometer Gravitational-wave Observatory (LIGO)[7,15] and the French–Italian detector, Advanced Virgo.[16,17] In addition, complementary space-based gravitational-wave detectors similar to the previously planned LISA[18] are expected within the next decades. Ongoing pulsar timing-based searches[19–21] for gravitational-waves can access signal frequencies below $\sim 10^{-7}$ Hz and are expected to produce results on new physics within the next decade.

Cosmic source population models imply[22] a detection rate for binary neutron star coalescence[23–25] between approximately 0.4 and 400 events annually for the Advanced gravitational-wave detector network. The possible rate of detections for eccentric binaries[26–28] was also predicted[13] to be above ~ 5 events annually with a single gravitational-wave detector signal-to-noise ratio of 5, depending on the model used.

Foreseeing the wide range of possibilities in gravitational-wave astronomy, the worldwide gravitational-wave community is taking steps[29] toward developing a baseline design for the third generation of interferometric gravitational-wave detectors (e.g., Einstein Telescope [ET][14,30–33]). The target sensitivity—representing an order of magnitude increase in cosmic distance reach for signal frequencies near 100 Hz compared to the Advanced detectors currently under construction—would open the way to an era of routine detections for an extended range of gravitational-wave emitting sources.

Experiments searching for gravitational-waves will enable new scientific possibilities[13,14,22,26,34,35] especially with the anticipated advent of regular, direct detections. An Advanced gravitational-wave detector network, encompassing several continents, is expected to collect data at unprecedented sensitivities in the coming decades. In addition to the United States and European detector network, the Japanese Large Cryogenic Gravitational-wave Telescope (LCGT)[36,37] is expected to join the network in the fourth quarter of this decade. Currently, the gravitational-wave community is actively surveying possibilities to move one of the three Advanced LIGO detectors to an overseas location such as India. Additional detectors in Australia and/or Argentina would be very valuable and could be contemplated at a later date.[38–40]

Comprehensive multimessenger astrophysics can be thought of as a generalization of multiwavelength astronomy because it connects a wide range of distinct observations of the same astrophysical event or system. In a broad sense, it can include data from γ-ray, X-ray, ultraviolet, optical, infrared, radio, gravitational-wave,[6] and neutrino observatories or subset of them.[14,41–44] Electromagnetic observations of astrophysical sources with plausible associated gravitational-wave emission have already initiated searches for gravitational-waves (see e.g., Refs. 45–50). The supplementary information[51] (specific trigger time, expected frequency range, direction, progenitor information, etc.) embedded in the collected electromagnetic data allows for the mining of gravitational-wave data closer to the ultimate limit, the gravitational-wave detector's noise floor level. It also allows for observational statements[46,50] that are relevant to the astrophysics community and might not have been accessible to gravitational-wave only searches.[52–55] For example, in many cases the source distance and estimates of the energy reservoir of the progenitor are also available, making it possible to propagate the energetics of the observation or nonobservation by the gravitational-wave detector to the astrophysical source itself and to compare the gravitational-wave energy to the energy emitted in electromagnetic radiation.[14,41,47,50,56] Furthermore, multimessenger detections of the future, encompassing gravitational-wave observations, could enable us to solve great cosmic puzzles without crucial missing pieces.

During the past decade, multimessenger observations using gravitational-waves focused on searches that were initiated[46,50,51] and guided by nongravitational-wave related observations (such as γ-rays) to constrain the analysis of gravitational-wave data. The recent emergence of a worldwide network of gravitational-wave observatories enabled the reconstruction of the sky position[39,57,58] of the gravitational-wave source candidates. Consequently, low-latency data analysis pipelines[59,60] allowed the gravitational-wave and astronomy communities to pioneer the inverse approach of electromagnetic follow-up[61] to gravitational-wave event candidates, initiating multimessenger observations in the hope of recording apparent optical and X-ray afterglows. In 2010 observation requests had been sent to wide-field optical telescopes and other instruments, such as QUEST, TAROT, ZADKO, Pi of the Sky, ROTSE, SkyMapper, the Palomar Transient Factory, and the Swift satellite. Low-latency electromagnetic follow-up observations of gravitational-wave event candidates[59,60] were initiated at these observatories to identify possible optical counterparts.[43,56,62–65] As an additional approach and opening toward particle type messengers, collaborations have been established between gravitational-wave telescopes and high-energy neutrino observatories—IceCube and ANTARES—to exploit the joint, early detection[42,66] of common cosmic sources, some of which might be difficult to observe through an electromagnetic instrument.

In this paper, the section titled "Birth of gravitational-wave multimessenger astronomy" provides a glimpse of some of the past results from the transient gravitational-wave field; in the section titled "Challenges and open questions in the future of gravitational-waves," I give an optimistic outlook on some aspects of the future of transient multimessenger science with the worldwide network of gravitational-wave observatories; and the section titled "Emerging opportunities in multidisciplinary science" provides a very personal example of exciting and synergistic multidisciplinary research directions in New York that was enabled by decades of experience in fundamental physics research.

Birth of gravitational-wave multimessenger astronomy

During the past decade multimessenger searches for gravitational-waves[42,43,45–51,61,67–74] were established. Examples of selected results are highlighted in this section.

Gamma-ray bursts (GRBs) were one of the first targets of multimessenger[51] searches using gravitational-waves. GRBs are energetic flashes of γ-rays often tied to black hole formation processes.[75–77] GRBs are typically categorized by their characteristic duration.[78,79] Long GRBs have durations of $\gtrsim 2$ sec[80–82] and several long GRBs have been spatially and temporally coincident with supernovae.[83–87] Most short GRBs (durations $\lesssim 2$ sec)

are associated with distant galaxies, however many powerful short bursts of γ-rays were also observed from sources in the Milky Way, such as the soft γ repeater SGR 1806–20.[88–91] Soft Gamma Repeaters (SGRs) are possibly highly magnetized neutron stars observed to emit short-duration X- and γ-ray bursts intermittently. In rare events, these sources have exhibited giant flares that can reach peak electromagnetic luminosities of up to 10^{47} erg/sec.[92] However, the leading hypothesis posits that most short extragalactic GRBs come from the merging of neutron star or neutron star + black hole binaries.[93,94] Such binary systems are also expected to be strong gravitational-wave emitters at frequencies accessible to ground-based gravitational-wave detectors.[73,95–101] To date no observations have definitively confirmed the association between short GRBs and binary mergers; such confirmation is expected from direct detection of gravitational-waves in coincidence with short GRBs.

During the past approximately eight years, the LIGO and Virgo Collaborations collected data coincident with hundreds of γ-ray bursts detected by Swift, Fermi, HETE, IINTEGRAL, and other satellites. As a result, numerous observational upper limits on gravitational-wave emission have been published.[42,43,45–51,61,67–74,102,103] In one use of LIGO data, the short-hard γ ray burst GRB070201—possibly associated with the Andromeda galaxy (M31)—triggered a gravitational-wave search[50] that proved significant from both an astronomical and a sociological viewpoint. A binary compact object merger in M31, which is $\cong 770$ kpc from Earth, would have been detectable by the observing gravitational-wave observatories, yet no gravitational-waves were found in the analysis of LIGO data. Therefore any hypothetical inspiraling binary source had to have been located well beyond M31. The alternative, an extragalactic SGR giant flare progenitor in M31, remains the favored possibility[104–108] and would be an interesting discovery itself.

Multimessenger searches were also pursued for Galactic SGRs[109] situated within $\simeq 15$ kpc.[88,90,109] It is theorized that highly energetic γ-ray emission from SGRs[110–114] or solid quark stars[115–117] could also be accompanied by emission of gravitational-waves.[113,114,117] In addition, quasi-periodic oscillations (QPOs)[118–121] electromagnetically embedded for hundreds of seconds in the pulsating X-ray tail of SGR1806-20[118] and SGR1900+14 are also interesting as their well-characterized frequencies lie in the 30–2,000 Hz frequency range,[118–120] coinciding with the sensitive bandwidth of ground-based gravitational-wave detectors. Some of the plausible interpretations attributed the observed QPOs to seismic modes of the neutron star and thus they may also be accompanied by gravitational-wave[47] emission. The most stringent upper limit result[47] on the energy emitted in the form of gravitational-waves during an observed QPO event corresponds to the order of the total isotropic energy observed to be emitted in the form of electromagnetic waves from SGR giant flares.

The search for gravitational-waves associated with the initial flare phase of SGR bursts observed by satellites during data collection periods of ground-based interferometric gravitational-wave detectors[46,49,122] provided results that sparked interest in SGR modeling and theory.[114,123] During the Advanced gravitational-wave detector era, the energy scale probed by the gravitational-wave data analysis could be several orders of magnitude below the energy scale of the electromagnetic emission in giant flares, deep in the intriguing region of previously unexplored gravitational-wave emission energy scales.[114–117]

High-energy neutrinos (HEN) are also targets for multimessenger searches for gravitational-waves. Together with gravitational-waves, they could be emitted from GRBs,[77,124–127] some of which are likely to be associated with neutron star–neutron star or neutron star–black hole mergers.[93,94,128,129] There are several other putative common sources of gravitational-waves and HENs, such as core-collapse supernovae, SGRs, and microquasars.[14,130] Interestingly, the GW+HEN multimessenger search for joint sources can be sensitive to progenitor events that have little or no detectable electromagnetic emission. Coincident[131] observations of gravitational-waves and neutrinos from GRBs could improve our understanding of the details of the astrophysical processes connecting the gravitational collapse/merger of compact objects and black hole formation. The pioneering international collaboration between the ANTARES, IceCube, LIGO, GEO600, and Virgo Collaborations on the joint analysis[42,132,133] of gravitational-wave and HEN data was a breakthrough on the multimessenger front. The astronomical reach of IceCube[134] and ANTARES[135] neutrino observatories in the high-energy regime, extending to extragalactic distances, offers a complementary sky coverage, and allows the energy, arrival time, and direction of individual neutrinos to be determined. The Advanced gravitational-wave detector network shall also provide extended extragalactic reach as well as source localization information comparable to the neutrino detectors. The scientific reach of current and planned experiments to multimessenger sources of gravitational-waves and HEN was recently explored in Ref. 66. Constraints on the rate of multimessenger transients were derived based on independent observations by the initial LIGO and Virgo gravitational-wave detectors and the partially completed IceCube (40-string) HEN detector, taking into account the blue-luminosity-weighted distribution of nearby galaxies. The results indicate that future observations may be able to exclude and constrain various emission models for the most interesting source classes, just as is also possible with many of the other multimessenger observations. Some of these future observations are discussed in the next section.

Challenges and open questions in the future of gravitational-waves

In the coming decades, more sensitive instruments capable of observing cosmic gravitational-waves will be built (see e.g., Fig. 1), enabling unprecedented scientific reach. Technical improvements brought by these second and third generation gravitational-wave detectors and the creation of worldwide detector networks can lead to the realization of routine detection, increased signal frequency coverage, and increased cosmic range coverage. These in turn allow us to further our understanding of the internal astrophysical mechanisms of stellar explosions, add to our understanding of gravity and general relativity, probe into the early universe, and attempt to answer other open questions in astrophysics and cosmology. For example, the prospect of finding out what really happens in a core-collapse supernova before the light and neutrinos escape and measuring the delay in between neutrinos and gravitational-waves in a core collapse even[137] is quite captivating.

Analysis of data collected by second and third generation worldwide interferometric gravitational-wave detector networks shall exploit synergistic astrophysics relying on a comprehensive range

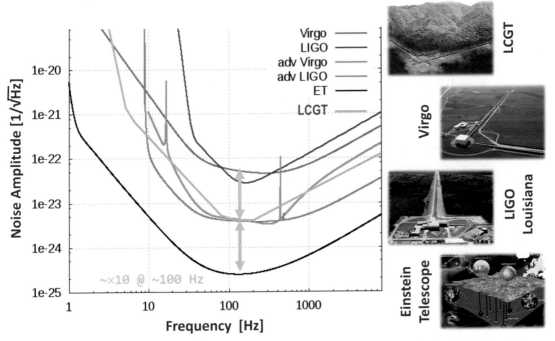

Figure 1. Nominal noise curves of past and future interferometric gravitational-wave detectors. The area above the curves is open for exploration. The vertical axis represents the strainnoise[136] of the detectors, while the horizontal axis represents the frequency. The inset on the right shows the aerial view of some of the existing detector infrastructures and the conceptual view of LCGT and ET. Data and images are courtesy of VIRGO Collaboration, LCGT Collaboration, LIGO Laboratory, LIGO Scientific Collaboration and ET Science Team.

of observations, including detected waveforms in the gravitational-wave spectrum, as well as γ-rays, X-rays, radio, optical, and astrophysical low-energy neutrinos and HEN. For example, only the possibility that cosmic sources of γ-ray bursts and core-collapses that are invisible for the electromagnetic observer[138–140] might be found is exhilarating. The addition of precise gravitational-wave information to the multimessenger approach can allow the field to crack some of the toughest cosmic puzzles of today and tomorrow.

Information about cosmic sources have already been deduced from nondetection results, but many of the exciting questions remain to be answered (see e.g., Refs. 41, 56, 64, 141). Gravitational-wave astrophysics is a frontier field and it will take several generations of ground-based gravitational-wave detectors to fully explore the broad range of cosmic mysteries explicable through gravitational-wave detections.[41,142] The time line of the operational status of the planned and active gravitational-wave detectors is outlined in the Roadmap of

the Gravitational Wave International Committee (GWIC),[143] which envisions groundbreaking instrumental and scientific advances. The current instrumentation effort focuses on the commissioning of Advanced LIGO,[15,144] Advanced VIRGO,[145] and LCGT[37] detectors. These second generation gravitational-wave detectors are aiming for more than an order of magnitude improvement in sensitivity as well as expanded gravitational-wave signal frequency coverage over initial detectors. This will result in an increase in cosmic volume coverage of over three orders of magnitude.[146] Thanks to these advances a half a day's worth of data collected by an Advanced ground-based detector network might contain more astrophysical information than all of the gravitational-wave data collected before. Their observations could revolutionize astronomy.

It is feasible but not necessarily assured that the direct detection of gravitational-waves could be made within the decade using an Advanced detector network. To prepare for the routine detection era that shall follow the first direct detection of gravitational-waves, significant collaborative work among theorists, phenomenologists, and observationalists will be needed in the near future. These shall enable the full exploitation of the observational[147,148] results to achieve in-depth understanding of the internal mechanisms and parameters leading to the cosmic explosions. It will allow the scientific community to answer long-standing puzzles of nature, in many cases through multimessenger gravitational-wave observations. The prospect of making fundamental advances including finally measuring the speed of gravitational-waves and gravity,[149,150] determining whether or not gravity violates parity conservation,[151,152] and probing whether Einstein's theory of general relativity remains valid in the strong-field regime of black holes[153] is both fascinating and revolutionary.

Detections will be analyzed in real time to extract information about the processes fueling violent cosmological events, such as binary mergers, GRBs, SGRs, or supernovae. Gravitational-wave signals will be used to trigger measurements of other messengers from a given direction by turning a careful selection of appropriate telescopes toward the anticipated direction of a multimessenger source. This will open a new era for optical astronomy, enabling the optical detection of sources in very early stages and giving access to populations that are difficult to observe otherwise.

Third generation gravitational-wave detectors—such as the Einstein Telescope concept[15,30])—are expected to have sensitivities about ten times better than the Advanced gravitational-wave detectors currently under construction. These detectors will be built underground to insulate the instrument from seismic disturbances and may feature three co-located interferometers placed in a triangular shape. One of the main advances during the era of third generation ground-based gravitational-wave detectors might be a vast increase in the number of gravitational-wave detections. Regular detection of a given gravitational-wave source type shall provide information unobtainable from the detection of a single event. The use of multiple high precision measurements to set statistics on the occurrence of observed sources, and map the distribution of gravitational-wave source parameters is clearly an attractive prospect. For example, detection of gravitational-waves during eccentric encounters of compact objects[13,26–28,154] of black holes would offer a unique observational probe for constraining the stellar black hole mass function of dense clusters and galactic cores. Third generation gravitational-wave detectors should also be able to measure previously unobserved physical parameters of GRBs and compact binary systems of black holes and neutron stars in detail. These measurements could constrain the equation-of-state of neutron stars and are expected to contribute to the solution of the enigma of γ-ray bursts. General relativity will be tested by comparing observations to our theoretical expectations, which might provide phenomenological constraints on alternative theories of gravity.

Third generation gravitational-wave detectors may observe sources at high redshifts and consequently provide important information about their populations, cosmological evolution, and large-scale structure. Gravitational-wave emission from compact binary mergers can reveal information about their parameters, including their distance and might answer the critical question: Do such gravitational-wave based measurements of distances agree with the concordance cosmology? Because these "standard gravitational-wave siren"[155–157] distance measurements can be complementary to the current cosmic-distance-ladder, they might also be used to precisely calibrate existing electromagnetic or other standard candles.[15,30,158,159] For example, if short GRBs emitted by compact binary mergers are jointly detected through electromagnetic and gravitational-wave observations then these complementary measurements could be used to independently measure the fundamental Hubble parameter, and possibly the dark matter and dark energy densities as well.

Detection of gravitational-wave signals originating from supernovae will finally become an expectation during the lifetime of the far-reaching third generation detector network. Joint detection of electromagnetic, neutrino, and gravitational-wave signals from core-collapse supernovae[137,160–164] is a widely anticipated type of synergistic observation[165–171] that is expected to provide new information on the supernova engine and the dynamics of matter within the source.[14,172]

It is likely that generations of gravitational-wave observatories with revolutionary scientific reach will observe the cosmos as a worldwide network. There will be rich and fruitful interaction among the gravitational-wave, electromagnetic, and neutrino observations that might make the next couple of decades the golden era of comprehensive multimessenger astronomy. Thus, it is fair to ask, what are we hoping to learn about the natural world?

In the coming decade(s), many outstanding, fundamental questions about the nature of gravity and elusive cosmic objects[14,29,31,41,56,142,143] might be answered. Among many others, the highlights include:

i. What is the central process driving γ-ray bursts?
ii. Low luminosity and "choked" GRBs were hypothesized, but do they really exist?[14]
iii. Can we map the shape of GRB jets from gravitational-waves?
iv. How many gravitational-wave polarizations are there?
v. Can we detect Shapiro time-delay in the lab for small masses?[173]
vi. Do "quark stars" really exist?
vii. Can we detect the nonaxisymmetric crust and core dynamics of magnetar flares?[113]
viii. How high is the tallest mountain and what is the structure of neutron stars?[123,174]
ix. How does anisotropic neutrino emission influence supernovae?[175]

Of course, the most exciting phenomenon discovered by gravitational-wave astrophysics might not be listed anywhere: the opening of a new window to our Universe may very well result in completely unanticipated discoveries.

This is a perfect point to finish the review of the results and inspiring prospects of large worldwide astrophysics collaborations, and examine the local opportunities enabled by decades of broad research experience in basic sciences. Although fundamental science efforts are extremely exciting and can often be as enjoyable as any refined form of art, scientists must also be keenly aware of the world closer to home. To emphasize this point, I will highlight some of the exciting biophysics directions we pursue at Columbia University in small collaborations with colleagues of diverse interest and expertise.

Emerging opportunities in multidisciplinary science

Newly developed concepts and advances triggered by modern physics are responsible for a grow-

ing number of contributions to the fields of biology, medicine, and technology. Magnetic resonance imaging and the World Wide Web (WWW) are probably the best known examples, along with positron emission tomography and computed tomography (CT). Some of these advances developed out of fundamental research, but many have also grown out of the desire of scientists to contribute to the betterment of humanity as a whole. The application of accumulated basic research experience from a broad range of sciences to areas where innovative research can yield a significant human impact is a noble and useful goal. Expertise in the physical sciences can be used to address pressing questions in our everyday lives as well as target difficult questions of pressing social concern. Among other examples, advances in understanding the connection between genetics, disease, and behavior can be gained through novel optics, new instrumentation, and advanced data analysis techniques. In the next three paragraphs, we sketch some of the interesting multidisciplinary directions we are exploring in our laboratories at Columbia.

The connection between behavior and genetics is one of the big questions of modern science amiable to major breakthroughs arising from expertise in physical sciences. For example, the ability to move through one's environment distinguishes members of the animal kingdom from other forms of life on Earth, and locomotion is essential for animal survival. However, despite the importance of locomotion to animal life, and the potential impact on medicine and society, our understanding of the neural circuitry involved is surprisingly limited. Deciphering the relevant neuronal circuitry that provides terrestrial animals with the ability to walk in a coordinated manner shall be a great step forward. BioOptics-based technology developed at Columbia University aims to identify neurons and genes that are required for motor coordination in the fruit fly model and could yield fundamental new insights into the biological mechanisms that govern motor coordination.

Neurodegenerative diseases represent a major treatment challenge to the biomedical research community as well as a human catastrophe on both the personal and global scale. These debilitating disorders directly and indirectly affect the lives of tens of millions of people and cause tremendous damage to the economy. Remarkably few effective interventions exist and yet the prevalence of these pathologies is in ascendency. The combination of automated assessments of gait, coordination, and somatosensation in an established animal model of a neurodegenerative disease directly permits the precise, quantitative, nonbiased, fast, and statistically rigorous assessment of new treatment modalities for a range of neurodegenerative diseases.

Arthropods have been evolving for millions of years to efficiently analyze and respond to their complex environments. They have developed highly sophisticated sensors, can easily navigate through difficult terrain, can recognize odor threads, heat sources, food, animals, humans and other desirable targets, and are capable of solving highly complex tasks far beyond the abilities of even the most powerful modern computers and robots. Consequently, arthropods can be regarded as low-cost, low-power, radiation-hardened, sophisticated, self-replicating navigation systems with multisensor signal processing, and neural-network-based computing capabilities. Integrating arthropod neural systems with machines can add to basic science and create innovative applications that save lives.

Astrophysics deals with extremely remote and complex phenomena seemingly far removed, for example, from the study of insect nervous systems. Nonetheless, there is a thread from the study of enigmatic cosmic phenomena to that of fundamental processes in neural systems. Understanding the neural architecture and design of neuromorphic techniques will aid us in both the biological and physical sciences. Research on sensors, behavior, and nervous systems may well lead us toward better sensing, control, and analysis tools in technology and astrophysics, thus completing the circle of scientific discovery.

Acknowledgments

The Columbia Experimental Gravity group is grateful for the generous support from Columbia University in the City of New York, from the National Science Foundation under cooperative agreement PHY-0847182, and the Bill and Melinda Gates Foundation. The author would like to thank his excellent collaborators, especially Z. Márka, I. Bartos, S. Sturley and R. Mann, M. Tse, D. Murphy, R. Frey, D. Shoemaker, S. Whitcomb, and J. Kanner for their invaluable comments on the manuscripts of this paper. The author is grateful to hundreds of colleagues for making gravitational-wave experiments and results a reality. The review, analysis, opinions, and projections expressed in this paper are solely the author's and do not necessarily reflect the opinion of large collaborations involved in the research. This paper has been assigned LIGO Document Number LIGO-P1100139.

Conflicts of interest

The author declares no conflicts of interest.

References

1. Rees, M.J. 1973. Astrophysical aspects of gravitational waves. In *Sixth Texas Symposium on Relativistic Astrophysics*. D.J. Hegyi, Ed. *Ann. N.Y. Acad. Sci.* **224:** 118.
2. Teukolsky, S.A. 1975. Gravitational waves and black holes. In "Seventh Texas Symposium on Relativistic Astrophysics." E.J. Fenyves, L. Motz & P.G. Bergmann, Eds. *Ann. N.Y. Acad. Sci.* **262:** 275–283.
3. Amaldi, E. 1980. Summary of workshop on gravitational radiation. In *Ninth Texas Symposium on Relativistic Astrophysics*. J. Ehlers, J.J. Perry & M. Walker, Eds. *Ann. N.Y. Acad. Sci.* **336:** 322–333.
4. Schutz, B.F. 1989. Gravitational radiation. *Ann. N.Y. Acad. Sci.* **571:** 27–43.
5. Thorne, K.S. 1995. Gravitational radiation. In "Seventeenth Texas Symposium on Relativistic Astrophysics and Cosmology." H. Böhringer, G.E. Morfill & J.E. Trümper, Eds. *Ann. N.Y. Acad. Sci.* **759:** 127–152.
6. Hughes, S.A., S. Marka, P.L. Bender & C.J. Hogan. 2001. In "Proceedings of the APS/DPF/DPB Summer Study on the Future of Particle Physics (Snowmass 2001)." N. Graf, Ed: eConf, C010630, P402, arXiv:astro-ph/0110349.
7. The LIGO Scientific Collaboration. 2009. LIGO: the Laser Interferometer Gravitational-Wave Observatory. *Rep. Prog. Phys.* **72:** 076901.
8. Abbott, B., R. Abbott, R. Adhikari, *et al.* 2004. Detector description and performance for the first coincidence observations between LIGO and GEO. *Nucl. Instrum. Meth. A.* **517:** 154–179. Preprint arXiv:gr-qc/0308043.
9. Deffayet, C. & K. Menou. 2007. Probing gravity with spacetime sirens. *Astrophys. J. Lett.* **668:** L143–L146. Preprint 0709.0003.
10. Desai, S., E.O. Kahya & R.P. Woodard. 2008. Reduced time delay for gravitational waves with dark matter emulators. *Phys. Rev. D.* **77:** 124041. Preprint 0804.3804.
11. Fregeau, J.M., S.L. Larson, M.C. Miller, *et al.* 2006. Observing IMBH-IMBH binary coalescences via gravitational radiation. *Astrophys. J. Lett.* **646:** L135–L138. Preprint arXiv:astro-ph/0605732.
12. Brown, D.A., J. Brink, H. Fang, *et al.* 2007. Prospects for detection of gravitational waves from intermediate-mass-ratio inspirals. *Phys. Rev. Lett.* **99:** 201102. Preprint arXiv:gr-qc/0612060.
13. O'Leary, R.M., B. Kocsis & A. Loeb. 2008. ArXiv e-print 0807.2638 **807**. Preprint 0807.2638.

14. Chassande Mottin, E., M. Hendry, P.J. Sutton & S. Márka. 2011. Multimessenger astronomy with the Einstein Telescope. *Gen. Relativ. Gravit.* **43:** 437–464. Preprint 1004.1964.

15. Harry, G.M., the LIGO Scientific Collaboration. 2010. Advanced LIGO: the next generation of gravitational wave detectors. *Class. Quant. Gravity* **27:** 084006.

16. Virgo Collaboration. Available at: https://wwwcascina.virgo.infn.it/advirgo/docs.html.

17. Accadia, T., F. Acernese, F. Antonucci, *et al.* 2011. Status of the Virgo project. *Class. Quant. Gravity* **28:** 114002.

18. Danzmann, K. 1995. LISA-Laser Interferometer Space Antenna for gravitational wave measurements. In "Seventeenth Texas Symposium on Relativistic Astrophysics and Cosmology." H. Böhringer, G.E. Morfill & J.E. Trümper, Eds. *Ann. N.Y. Acad. Sci.* **759:** 481–484.

19. Demorest, P., J. Lazio, A. Lommen, *et al.* 2009. Gravitational wave astronomy using pulsars: massive black hole mergers the early universe. In *astro2010: The Astronomy and Astrophysics Decadal Survey.* ArXiv Astrophysics e-prints p. 64. Preprint 0902.2968.

20. Verbiest, J.P.W., M. Bailes, W.A. Coles, *et al.* 2009. Timing stability of millisecond pulsars and prospects for gravitational-wave detection. *Mon. Not. Roy. Astron. Soc.* **400:** 951–968. Preprint 0908.0244.

21. Hobbs, G., A. Archibald, Z. Arzoumanian, *et al.* 2010. *Class. Quant. Gravity* **27:** 084013. Preprint 0911.5206.

22. Abadie, J., B.P. Abbott, R. Abbott, *et al.* 2010. *Class. Quant. Gravity* **27:** 173001. Preprint 1003.2480.

23. Abadie, J., B.P. Abbott, R. Abbott, *et al.* 2010. *Phys. Rev. D.* **82:** 102001. Preprint 1005.4655.

24. Abbott, B.P., R. Abbott, R. Adhikari, *et al.* 2009. *Phys. Rev. D.* **80:** 047101. Preprint 0905.3710.

25. Abbott, B.P., R. Abbott, R. Adhikari, *et al.* 2009. *Phys. Rev. D.* **79:** 122001. Preprint 0901.0302.

26. Kocsis, B., M.E. Gáspár & S. Márka. 2006. *Astrophys. J.* **648:** 411–429. Preprint arXiv:astro-ph/0603441.

27. Washik, M.C., J. Healy, F. Herrmann, *et al.* 2008. ArXiv e-print 0802.2520 **802.** Preprint 0802.2520.

28. Kocsis, B. & J. Levin. 2011. ArXiv e-prints. Preprint 1109.4170.

29. Whitcomb, S.E. *et al.* 2009. "*NAS WEB.*" Available at: https://dcc.ligo.org/cgi-bin/DocDB/ShowDocument?docid=1587.

30. Team, E.S. 2011. Einstein gravitational wave telescope conceptual design study. Available at: https://tds.ego-gw.it/itf/tds/file.php?callFile=ET-0106C-10.pdf.

31. Sathyaprakash, B.S., B.F. Schutz & C. Van Den Broeck. 2010. *Class. Quan. Gravity* **27:** 215006. Preprint 0906.4151.

32. Hild, S., M. Abernathy, F. Acernese, *et al.* 2011. *Class. Quan. Gravity* **28:** 094013. Preprint 1012.0908.

33. Punturo, M., M. Abernathy, F. Acernese, *et al.* 2010. *Class. Quant. Gravity* **27:** 084007.

34. Kalogera, V. *et al.* 2004. The cosmic coalescence rates for double neutron star binaries. *Astrophys. J.* **601:** L179–L182. Erratum-ibid. **614** (2004) L137.

35. Abbott, B.P., R. Abbott, M. Abernathy, *et al.* 2011. ArXiv e-prints. Preprint 1109.1809.

36. Kuroda, K., LCGT Collaboration. 2010. *Class. Quant. Gravity* **27:** 084004.

37. Arai, K., R. Takahashi, D. Tatsumi, *et al.* & The LCGT Collaboration. 2009. *Class. Quant. Gravity* **26:** 204020.

38. Barriga, P., D.G. Blair, D. Coward, *et al.* 2010. *Class. Quant. Gravity* **27:** 084005.

39. Fairhurst, S. 2011. *Class. Quant. Gravity* **28:** 105021. Preprint 1010.6192.

40. LIGO Scientific Collaboration. 2010. Report of the committee to compare the scientific cases for two gravitational-wave detector networks: (ahlv) australia, hanford, livingston, virgo; and (hhlv) two detectors at hanford, one at livingston, and virgo. Available at: https://dcc.ligo.org/cgi-bin/DocDB/ShowDocument?docid=11604.

41. Márka, S., the Ligo Scientific Collaboration & the Virgo Collaboration. 2010. *J. Phys. Conf. Ser.* **243:** 012001.

42. Aso, Y., Z. Márka, C. Finley, *et al.* 2008. *Class. Quant. Gravity* **25:** 114039. Preprint 0711.0107.

43. Sylvestre, J. 2003. *Astrophys. J.* **591:** 1152–1156. Preprint arXiv:astro-ph/0303512.

44. Stubbs, C.W. 2008. *Class. Quant. Grav.* **25:** 184033. Preprint arXiv:0712.2598.

45. Acernese, F., M. Alshourbagy, P. Amico, *et al.* 2008. *Class. Quant. Gravity* **25:** 225001–+ (Preprint 0803.0376).

46. Abbott, B., R. Abbott, R. Adhikari, *et al.* 2008. *Phys. Rev. Lett.* **101:** 211102. Preprint 0808.2050.

47. Abbott, B., R. Abbott, R. Adhikari, *et al.* 2007. *Phys. Rev. D.* **76:** 062003. Preprint arXiv:astro-ph/0703419.

48. Abbott, B., R. Abbott, R. Adhikari, *et al.* 2009. *Astrophys. J. Lett.* **701:** L68–L74. Preprint 0905.0005.

49. Abadie, J., B.P. Abbott, R. Abbott, *et al.* 2011. *Astrophys. J. Lett.* **734:** L35. Preprint 1011.4079.

50. Abbott, B., R. Abbott, R. Adhikari, *et al.* 2008. *Astrophys. J.* **681:** 1419–1430. Preprint 0711.1163.

51. Abbott, B., R. Abbott, R. Adhikari, *et al.* 2008. *Class. Quant. Gravity* **25:** 114051. Preprint 0802.4320.

52. The LIGO Scientific Collaboration. 2007. *Class. Quant. Gravity* **24:** 5343–5369. Preprint arXiv:0704.0943.

53. The LIGO Scientific Collaboration. 2008. *Phys. Rev. D.* **77:** 062002.

54. Abbott, B.P., R. Abbott, R. Adhikari, *et al.* 2009. *Phys. Rev. D.* **80:** 102001. Preprint 0905.0020.

55. Abbott, B., R. Abbott, R. Adhikari, *et al.* 2008. *Class. Quant. Gravity* **25:** 245008. Preprint 0807.2834.

56. Bloom, J.S. *et al.* 2009. Astro2010 Decadal Survey Whitepaper: Coordinated Science in the Gravitational and Electromagnetic Skies. Preprint arXiv: 0902.1527.

57. Nissanke, S., J. Sievers, N. Dalal & D. Holz. 2011. *Astrophys. J.* **739:** 99. Preprint 1105.3184.

58. Klimenko, S., G. Vedovato, M. Drago, *et al.* 2011. *Phys. Rev. D.* **83:** 102001. Preprint 1101.5408.

59. Jameson, G. 2011. Imessenger Astronomy with Low-Latency Searches for Transient Gravitational Waves. PhD Thesis, Columbia University.

60. The LIGO Scientific Collaboration, Virgo Collaboration: Abadie, J., B.P. Abbott, R. Abbott, *et al.* 2011. ArXiv e-prints. Preprint 1109.3498.

61. Kanner, J. *et al.* 2008. ArXiv e-print 0803.0312 **803.** Preprint 0803.0312.

62. Hansen, B.M.S. & M. Lyutikov. 2001. *Mon. Not. Roy. Astron. Soc.* **322:** 695–701. Preprint arXiv:astro-ph/0003218.

63. Stubbs, C.W. 2008. *Class. Quant. Gravity* **25:** 184033. Preprint 0712.2598.

64. Phinney, E.S. 2009. Finding and using electromagnetic counterparts of gravitational wave sources. In *astro2010: The Astronomy and Astrophysics Decadal Survey* (*Astronomy vol 2010*) 235. Preprint 0903.0098.

65. Kulkarni, S. & M.M. Kasliwal. 2009. Transients in the local universe. In *Astrophysics with All-Sky X-Ray Observations*. N. Kawai, T. Mihara, M. Kohama & M. Suzuki, Eds.: 312. Preprint 0903.0218.

66. Bartos, I., C. Finley, A. Corsi & S. Márka. 2011. Observational constraints on multimessenger sources of gravitational waves and high-energy neutrinos. *Phys. Rev. Lett.* **107:**(25) 251101. Available at: http://link.aps.org/doi/10.1103/PhysRevLett.107.251101.

67. Acernese, F., P. Amico, M. Alshourbagy, *et al.* 2007. *Class. Quant. Gravity* **24:** 671.

68. Kalmus, P. *et al.* 2007. *Class. Quant. Gravity* **24:** 659.

69. Luca Matone, S. Marka. 2007. *Class. Quant. Gravity* **24:** 649.

70. Raffai, P., Z. Frei, Z. Márka & S. Márka. 2007. *Class. Quant. Gravity* **24:** 457.

71. Finn, L. S., S. D. Mohanty & J. D. Romano. 1999. *Phys. Rev. D.* **60:** 121101. Preprint arXiv:gr-qc/9903101.

72. Mohanty, S. D., S. Marka, R. Rahkola, *et al.* 2004. *Class. Quant. Gravity* **21:** 765.

73. Abbott, B., R. Abbott, R. Adhikari, *et al.* 2005. *Phys. Rev. D.* **72:** 042002. Preprint arXiv:gr-qc/0501068.

74. Abbott, B., R. Abbott, R. Adhikari, *et al.* 2008. *Phys. Rev. D.* **77:** 062004. Preprint 0709.0766.

75. Klebesadel, R. W., I. B. Strong & R. A. Olson. 1973. *Astrophys. J. Lett.* **182:** L85.

76. Meszaros, P. 2002. *Ann. Rev. Astron. Astrophys.* **40:** 137–169.

77. Piran, T. 2005. *Rev. Modern Phys.* **76:** 1143–1210. Preprint arXiv:astro-ph/0405503.

78. Kouveliotou, C. *et al.* 1993. *Astrophys. J. Lett.* **413:** L101–L104.

79. Gehrels, N. *et al.* 2006. *Nature* **444:** 1044–1046.

80. Kawai, N. *et al.* 2006. *Nature* **440:** 184–186.

81. Watson, D., J. N. Reeves, J. Hjorth, *et al.* 2006. *Astrophys. J. Lett.* **637:** L69–L72.

82. Jakobsson, P. *et al.* 2006. *Astron. Astrophys.* **447:** 897–903.

83. Campana, S. *et al.* 2006. *Nature* **442:** 1008–1010. Preprint arXiv:astro-ph/0603279.

84. Malesani, D. *et al.* 2004. *Astrophys. J. Lett.* **609:** L5–L8. Preprint arXiv:astro-ph/0405449.

85. Hjorth, J. *et al.* 2003. *Nature* **423:** 847–850 arXiv:astro-ph/0306347.

86. Galama, T.J. *et al.* 1998. *Nature* **395:** 670. Preprint arXiv:astro-ph/9806175.

87. Woosley, S.E. & J.S. Bloom. 2006. *Ann. Rev. Astron. Astrophys.* **44:** 507–556.

88. Hurley, K., C. Kouveliotou, T. Cline, *et al.* 1999. *Astrophys. J. Lett.* **523:** L37–L40. Preprint arXiv:astro-ph/9906020.

89. Nakar, E., A. Gal-Yam, T. Piran & D.B. Fox. 2006. *Astrophys. J.* **640:** 849–853.

90. Hurley, K., S.E. Boggs, D.M. Smith, *et al.* 2005. *Nature* **434:** 1098–1103. Preprint arXiv:astro-ph/0502329.

91. Palmer, D.M. *et al.* 2005. *Nature* **434:** 1107–1109.

92. Woods, P.M., C. Kouveliotou, M.H. Finger, *et al.* 2007. *Astrophys. J.* **654:** 470–486.

93. Bloom, J.S. *et al.* 2007. *Astrophys. J.* **654:** 878–884. Preprint arXiv:astro-ph/0607223.

94. Nakar, E. 2007. *Phys. Rep.* **442:** 166–236. Preprint arXiv:astro-ph/0701748.

95. Acernese, F. *et al.* 2002. *Class. Quant. Gravity* **19:** 1421–1428.

96. Willke, B., *et al.* 2002. *Class. Quant. Gravity* **19:** 1377–1387.

97. Kochanek, C.S. & T. Piran. 1993. *Astrophys. J. Lett.* **417:** L17.

98. The LIGO Scientific Collaboration. 2006. *Class. Quant. Gravity* **23:** 29–39.

99. The LIGO Scientific Collaboration. 2005. *Phys. Rev. D.* **72:** 082001. Preprint web location:gr-qc/0505041.

100. Finn, L.S., B. Krishnan & P.J. Sutton. 2004. *Astrophys. J.* **607:** 384–390.

101. Leonor, I., P.J. Sutton, R. Frey, *et al.* 2009. *Class. Quant. Gravity* **26:** 204017.

102. Abbott, B.P., R. Abbott, F. Acernese, *et al.* 2010. *Astrophys. J.* **715:** 1438–1452. Preprint 0908.3824.

103. Abadie, J., B.P. Abbott, R. Abbott, *et al.* 2010. *Astrophys. J.* **715:** 1453–1461. Preprint 1001.0165.

104. Ofek, E.O., M. Muno, R. Quimby, *et al.* 2008. *Astrophys. J.* **681:** 1464–1469. Preprint arXiv:0712.3585.

105. Mazets, E.P., R.L. Aptekar, T.L. Cline, *et al.* 2008. *Astrophys. J.* **680:** 545–549. Preprint arXiv:0712.1502.

106. Dolan, S.R. 2008. ArXiv e-print 0801.3805 **801.** Preprint 0801.3805.

107. Desai, S., E.O. Kahya & R.P. Woodard. 2008. *Phys. Rev. D.* **77:** 124041.

108. Kahya, E.O. 2008. ArXiv e-print 0801.1984. **801.** Preprint 0801.1984.

109. Esposito, P., G.L. Israel, S. Zane, *et al.* 2008. ArXiv e-print 0807.1658. **807.** Preprint 0807.1658.

110. Thompson, C. & R.C. Duncan. 1995. *Mon. Not. Roy. Astron. Soc.* **275:** 255–300.

111. Schwartz, S.J., S. Zane, R.J. Wilson, *et al.* 2005. *Astrophys. J. Lett.* **627:** L129–L132. Preprint arXiv:astro-ph/0504056.

112. Andersson, N. & K.D. Kokkotas. 1998. *Mon. Not. Roy. Astron. Soc.* **299:** 1059–1068. Preprint arXiv:gr-qc/9711088.

113. de Freitas Pacheco, J.A. 1998. *Astron. Astrophys.* **336:** 397–401. Preprint arXiv:astro-ph/9805321.

114. Ioka, K. 2001. *Mon. Not. Roy. Astron. Soc.* **327:** 639–662. Preprint arXiv:astro-ph/0009327.

115. Xu, R.X. 2003. *Astrophys. J. Lett.* **596:** L59–L62. Preprint astro-ph/0302165.

116. Owen, B.J. 2005. *Phys. Rev. Lett.* **95:** 211101. Preprint astro-ph/0503399.

117. Horvath, J.E. 2005. *Modern Phys. Lett. A* **20:** 2799–2804. Preprint astro-ph/0508223.

118. Israel, G.L., T. Belloni, L. Stella, *et al.* 2005. *Astrophys. J. Lett.* **628:** L53–L56. Preprint arXiv:astro-ph/0505255.

119. Strohmayer, T.E. & A.L. Watts. 2006. *Astrophys. J.* **653:** 593–601.

120. Watts, A.L. & T.E. Strohmayer. 2006. *Astrophys. J.* **637:** L117–L120.

121. Strohmayer, T.E. & A.L. Watts. 2005. *Astrophys. J.* **632:** L111–L114.

122. Kalmus, P. 2008. Search for Gravitational Wave Bursts from Soft Gamma Repeaters. PhD Thesis, Columbia University, City of New York.

123. Corsi, A. & B.J. Owen. 2011. *Phys. Rev. D.* **83:** 104014. Preprint 1102.3421.

124. Waxman, E. & J. Bahcall. 1997. *Phys. Rev. Lett.* **78:** 2292–2295. Preprint arXiv:astro-ph/9701231.

125. Vietri, M. 1998. *Phys. Rev. Lett.* **80:** 3690–3693. Preprint arXiv:astro-ph/9802241.

126. Waxman, E. 2000. *Phys. Script. Vol. T* **85:** 117. Preprint arXiv:astro-ph/9911395.

127. Kotake, K., K. Sato & K. Takahashi. 2006. *Rep. Prog. Phys.* **69:** 971–1143. Preprint arXiv:astro-ph/0509456.

128. Lee, W.H. & E. Ramirez-Ruiz. 2007. *New J. Phys.* **9:** 17. Preprint arXiv:astro-ph/0701874.

129. Sekiguchi, Y., K. Kiuchi, K. Kyutoku & M. Shibata. 2011. *Phys. Rev. Lett.* **107:** 051102. Preprint 1105.2125.

130. Pradier, T. 2009. *Nucl. Inst. Methods Phys. Res. A* **602:** 268–274. Preprint 0807.2562.

131. Baret, B., I. Bartos, B. Bouhou, *et al.* 2011. *Astropart. Phys.* **35:** 1–7. Preprint 1101.4669.

132. Aso, Y., C. Finley, Z. Marka, *et al.* 2008. Analysis method to search for coincidence events between the LIGO-Virgo Gravitational-wave Detector Network and the IceCube Neutrino Detector. *APS April Meeting and HEDP/HEDLA Meeting.* St. Louis, MI. Volume 53, Number 5, Abstracts.

133. Aso, S. Ando. van Elewyck V., Y., *et al.* 2009. *Int. J. Modern Phys. D* **18:** 1655–1659. Preprint 0906.4957.

134. IceCube. Available at: http://icecube.wisc.edu.

135. ANTARES. Available at: http://antares.in2p3.fr/.

136. Team, A.L. 2011. Advanced ligo reference design. Available at: https://dcc.ligo.org/cgi-bin/DocDB/Show Document?docid=1507.

137. Pagliaroli, G., F. Vissani, E. Coccia & W. Fulgione. 2009. *Phys. Rev. Lett.* **103:** 031102. Preprint 0903.1191.

138. Fischer, T., S.C. Whitehouse, A. Mezzacappa, *et al.* 2009. *Astron. Astrophys.* **499:** 1–15. Preprint 0809.5129.

139. Sumiyoshi, K., C. Ishizuka, A. Ohnishi, *et al.* 2009. *Astrophys. J. Lett.* **690:** L43–L46. Preprint 0811.4237.

140. O'Connor, E. & C.D. Ott. 2010. *Class. Quant. Gravity* **27:** 114103. Preprint 0912.2393.

141. Kulkarni, S.R. & M.M. Kasliwal. 2009. Astro2010 Decadal Survey Whitepaper: Transients in the local Universe. Preprint arXiv:0903.0218.

142. Sathyaprakash, B.S. & B.F. Schutz. 2009. *Living Rev. Relat.* **12:** 2. Preprint 0903.0338.

143. The Gravitational Waves International Committee. 2010. The future of gravitational wave astronomy. Available at: http://gwic.ligo.org/roadmap.

144. Shoemaker, D. 2009. Advanced LIGO anticipated sensitivity curves Tech. Rep. T0900288-v2 LIGO. Available at: https://dcc.ligo.org/cgi-bin/DocDB/ShowDocument? docid=2974.

145. VIRGO. Available at: http://wwwcascina.virgo.infn.it/advirgo/.

146. Hughes, S.A. 2003. *Ann. Phys.* **303:** 142–178. Preprint arXiv:astro-ph/0210481.

147. Seaman, R., R. Williams, A. Allan, *et al.* 2011. Sky event reporting metadata Version 2.0. ArXiv e-prints. Preprint 1110.0523.

148. Márka, S., for the LIGO Scientific Collaboration and the Virgo Collaboration. 2011. *Class. Quant. Gravity* **28:** 114013.

149. Dubovsky, S.L., P.G. Tinyakov & I.I. Tkachev. 2005. *Phys. Rev. Lett.* **94:** 181102. Preprint arXiv:hep-th/0411158.

150. Jacobson, T. & D. Mattingly. 2004. *Phys. Rev. D.* **70:** 024003. Preprint arXiv:gr-qc/0402005.

151. Yunes, N., R. O'Shaughnessy, B.J. Owen & S. Alexander. 2010. *Phys. Rev. D.* **82:** 064017. Preprint 1005.3310.

152. Alexander, S. & N. Yunes. 2009. *Phys. Rep.* **480:** 1–55. Preprint 0907.2562.

153. Kocsis, B., Z. Haiman & K. Menou. 2008. *Astrophys. J.* **684:** 870–887. Preprint 0712.1144.

154. Levin, J., S.T. McWilliams & H. Contreras. 2011. *Class. Quant. Gravity* **28:** 175001. Preprint 1009.2533.

155. Schutz, B.F. 1986. *Nature* **323:** 310.

156. Holz, D.E. & S.A. Hughes. 2005. *Astrophys. J.* **629:** 15–22. Preprint arXiv:astro-ph/0504616.

157. Nissanke, S., D.E. Holz, S.A. Hughes, *et al.* 2010. *Astrophys. J.* **725:** 496–514. Preprint 0904.1017.

158. Zhao, W., C. van den Broeck, D. Baskaran & T.G.F. Li. 2011. *Phys. Rev. D.* **83:** 023005. Preprint 1009.0206.

159. Chernoff, D.F. & L.S. Finn. 1993. *Astrophys. J. Lett.* **411:** L5–L8. Preprint arXiv:gr-qc/9304020.

160. Finn, L.S. 1991. *Ann. N. Y. Acad. Sci.* **631:** 156–172.

161. Arnaud, N., M. Barsuglia, M.A. Bizouard, *et al.* 2004. *Astropart. Phys.* **21:** 201–221. Preprint arXiv:gr-qc/0307101.

162. Janka, H.T., K. Langanke, A. Marek, *et al.* 2007. *Phys. Rep.* **442:** 38–74. Preprint arXiv:astro-ph/0612072.

163. Ott, C.D. 2009. *Class. Quant. Gravity* **26:** 204015. Preprint 0905.2797.

164. Leonor, I., L. Cadonati, E. Coccia, *et al.* 2010. *Class. Quant. Gravity* **27:** 084019. Preprint 1002.1511.

165. Scholberg, K. 2001. *Nucl. Phys. B Proc. Suppl.* **91:** 331–337. Preprint arXiv:hep-ex/0008044.

166. Fulgione, W. 2010. *J. Phys. Conf. Ser.* **203:** 012077.

167. Fukuda, S. *et al.* 2003. *Nucl. Inst. Methods Phys. Res. Sec. A: Accel., Spectrom., Detect. Assoc. Equip.* **501:** 418–462. ISSN 0168-9002. Available at: http://www.sciencedirect.com/science/article/B6TJM-4811KNK-B/2/efe9c5f7c6e3ae12e4764369e71910bf.

168. Ahrens, J., *et al.* 2004. *Astropart. Phys.* **20:** 507–532. ISSN 0927-6505. Available at: http://www.sciencedirect.com/science/article/B6TJ1-49W31VF-1/2/7d423e933a84316c3be00b1c193cbd77.

169. Nakamura, K. 2003. Hyper-kamiokande a next generation water cherenkov detector. In *Neutrinos and Implications for Physics Beyond the Standard Model.* R. Shrock, Ed. World Scientific Publishing Co.

170. Arnett, W.D., J.N. Bahcall, R.P. Kirshner & S.E. Woosley. 1989. *Annu. Rev. Astron. Astrophys.* **27:** 629–700.

171. Bethe, H.A. 1990. *Rev. Modern Phys.* **62:** 801–866.

172. Ott, C.D. 2009. *Class. Quantum Gravity* **26:** 063001. Preprint arXiv:0809.0695.

173. Ballmer, S., S. Márka & P. Shawhan. 2010. *Class. Quant. Gravity* **27:** 185018. Preprint 0905.0687.

174. Owen, B.J. 2005. *Phys. Rev. Lett.* **95:** 211101. Preprint arXiv:astro-ph/0503399.

175. Buonanno, A., G. Sigl, G.G. Raffelt, *et al.* 2005. *Phys. Rev. D.* **72:** 084001. Preprint arXiv:astro-ph/0412277.

Ann. N.Y. Acad. Sci. ISSN 0077-8923

ANNALS OF THE NEW YORK ACADEMY OF SCIENCES

Issue: *Blavatnik Awards for Young Scientists*

Harnessing the world's biodiversity data: promise and peril in ecological niche modeling of species distributions

Robert P. Anderson[1,2,3,4]

[1]Department of Biology, City College, [2]The Graduate Center, and [3]CREST Institute, The City University of New York, New York, New York. [4]Division of Vertebrate Zoology (Mammalogy), American Museum of Natural History, New York, New York

Address for correspondence: Robert P. Anderson, 526 Marshak Science Building, City College of the City University of New York, 160 Convent Avenue, New York, NY 10031. anderson@sci.ccny.cuny.edu

Recent advances allow harnessing enormous stores of biological and environmental data to model species niches and geographic distributions. Natural history museums hold specimens that represent the only information available for most species. Ecological niche models (sometimes termed *species distribution models*) combine such information with digital environmental data (especially climatic) to offer key insights for conservation biology, management of invasive species, zoonotic human diseases, and other pressing environmental problems. Five major pitfalls seriously hinder such research, especially for cross-space or cross-time uses: (1) incorrect taxonomic identifications; (2) lacking or inadequate databasing and georeferences; (3) effects of sampling bias across geography; (4) violation of assumptions related to selection of the study region; and (5) problems regarding model evaluation to identify optimal model complexity. Large-scale initiatives regarding data availability and quality, technological development, and capacity building should allow high-quality modeling on a scale commensurate with the enormous potential of and need for these techniques.

Keywords: biodiversity; climate change; ecological niche modeling; georeference; natural history museum; species distribution

Introduction

The relevance of museum data and ecological niche modeling to society

The vast majority of species on Earth remain known only from specimens housed in the research collections of natural history museums and herbaria (hereafter, "museums"), creating the need to infer from such information in order to estimate and characterize the planet's biodiversity.[1-6] The range, or geographic distribution, of a species constitutes one fundamental dimension of biodiversity—for many species, the only one feasible for study. Developed largely over the past two decades, computer techniques often termed *ecological niche modeling* (or *species distributional modeling*; see later) can harness information from museum specimens to model a species' environmental requirements and identify geographic areas suitable for it.[7] To do so, they use occurrence records of the species in conjunction with digital environmental data (especially regarding climate), typically interfacing with computer mapping software termed geographic infor-

mation systems (GIS). Despite the tremendous potential of such techniques and the explosion in their use over the past decade,[8] several factors have conspired to impede realization of their full potential. Based on my fieldwork and research in taxonomy, systematics, and ecological niche modeling, I offer this perspective regarding the enormous promise for, and substantial pitfalls involved in, transforming occurrence data from museums into niche models and their associated distributional predictions.

Here, I consider only modeling based on what are termed *presence-only* occurrence data, the principal data source available for studying species distributions (Table 1). Presence-only data, such as what can be gleaned directly from museum specimens, constitute records of places where a species has been documented—without any information regarding the species' abundance (abundance data) or indicating places where it does not occur (presence–absence data). For the most part, those latter data types exist only for very well-studied taxonomic groups in temperate areas of the world, especially Europe and North America. Hence, for most species,

doi: 10.1111/j.1749-6632.2011.06440.x

Table 1. Biodiversity information (definitions here and in other tables largely follow those of a recent theoretical and methodological treatment[7])

Term	Definition
Natural history museum	A museum with research collections and scientists studying biodiversity, often focused on animals. Frequently also contains extensive public exhibits.
Herbarium	A scientific institution with research collections and scientists studying biodiversity, focused on plants. Frequently also contains botanical gardens of living plants.
Taxonomy	The science of documenting and describing biological diversity, typically at the level of the species ("alpha taxonomy," including naming new species). *Revisionary taxonomy* refers to efforts to study a particular group (often a genus) in order to determine what species exist, describe any new to science, and characterize all of them (typically based on morphology, although increasingly leveraging molecular data as well).
Systematics	Often considered to include taxonomy, the science of characterizing biological diversity, including the evolutionary relationships among species and classifying them accordingly (typically via the Linnaean system, usually based on the results of phylogenetic analyses reconstructing evolutionary relationships, most commonly using molecular data).
GIS	Computer software that allows visualization of and calculations regarding spatial data.
Presence-only data	Data sets containing records of where a species has been observed to be present, but lacking any information regarding sites where it is absent.
Presence-background data	Data sets containing records of where a species has been observed to be present, as well as information regarding environmental variation across the study area (the "background") and whether or not sampling has occurred there (and if so, whether or not records of the species exist from those regions).
Presence-pseudoabsence data	Data sets containing records of where a species has been observed to be present, as well as sites where it has not been observed (but note that the species may actually inhabit these latter sites, which can lack records of it due to nonexistent or inadequate sampling).
Presence-absence data	Data sets containing records of where a species has been observed to be present, as well as sites where it is absent, or assumed to be, despite sampling efforts (but note that the species may actually inhabit these latter sites, if sampling is present but inadequate).
Abundance data	Data sets containing information regarding the abundance of a species at various sites.

and especially for poorly studied tropical regions of high biodiversity, presence-only data sets constitute the only information available.[4] Lacking absence data, most niche-modeling algorithms using presence-only occurrence data compare the environmental conditions of the sites that the species is known to inhabit with those in a sample of the "background" available in the study region.[9,a] Although other sources also can provide presence-only occurrence data, I focus on data from museums, which hold the only information available for the vast majority of species. I do not attempt a full summary of the field or its uses[8,10] or a theoretical synthesis (a void largely filled by a recent book,[7] whose terminology I follow). Rather, I aim to present this research area and its great utility to nonspecialists, point out the most overreaching and consistent issues that limit its use, and advocate for an updated vision regarding the mutualistic, but still vastly underdeveloped, relationship between museums and ecological niche modeling.[11]

Niche models based on presence-only occurrence data hold enormous utility in basic and applied biodiversity science, yet even the tremendous increase in the use of such techniques over the past decade pales in comparison with their staggering untapped potential (Table 2). In addition to exciting academic uses in biogeographic, ecological, and evolutionary studies, niche models hold great applied relevance—for example, in conservation biology, the mitigation of invasive species, and public health concerns related to zoonotic human diseases. The relevance of climatic change and other anthropogenic environmental alterations cuts across each of these areas. Largely because of society's need to forecast the future effects of our actions, it is the ability of niche models to predict after climatic change that has attracted the most attention to the field.[12,13] The power of these models to predict suitable conditions in other places and time periods derives from their niche-based nature. Instead of directly delineating in geography the places where a species occurs (its range, or occupied distributional area), they aim to characterize the environmental conditions suitable for the species (its niche; or more precisely, its ex-

isting fundamental Grinnellian niche). Subject to certain critical assumptions,[7] such a niche model then allows identification of the geographic areas that fulfill those requirements (its abiotically suitable area, termed the *potential distribution* by many workers) in the region or time period of interest.[b]

---

[b]The niche represents a central, unifying concept in ecology, albeit a complex one with varying perspectives and definitions presented over the past decades.[14–16] To place the ecological niche models (Table 2) discussed here in context, I follow the rearrangement of niche concepts explained at length in a recent theoretical and methodological treatment.[7] Although the division of niche concepts into complementary Grinnellian and Eltonian perspectives (that arguably constitute two ends of a continuum) espoused therein requires simplifying assumptions, this separation represents a highly useful distinction defensible in many circumstances. Under this framework, niche models using presence-only occurrence records (and typically a background sample of the study region) are based on environmental variables not affected by the presence of the species (termed *scenopoetic variables*). Generally most relevant and measured at coarse grains, such variables characterize the species' Grinnellian niche, delimiting density-independent factors that permit positive population growth rates for the species. Conveniently, this simplified perspective of the niche can be modeled as static sets of numbers. Such models differ from those of what has been termed the *Eltonian niche*[7] (detailed in a comprehensive treatment[14]), which considers variables modified by the presence of the species. Often relevant and measured at a fine spatial grain and small, local extent, these variables appear as density-dependent terms in population growth equations, requiring substantially more complicated mathematical formulations. Consideration of the Grinnellian niche via static sets of environmental conditions allows for definition of subsets of it that correspond to important biological situations, inspired by the long-standing concepts of fundamental and realized components of the niche (parallel to that distinction relevant for Eltonian niches[14]). For example, the full scenopoetic Grinnellian fundamental niche of a species likely contains various smaller subsets that represent conditions that can be termed the: existing fundamental niche, biotically reduced niche, invadable niche, and occupied niche. Researchers should consider such realities affecting the occurrence data for a species when planning a study using ecological niche modeling and when interpreting the resulting predictions. For example, dispersal limitations, biotic interactions, and even the limited set of environmental conditions existing on Earth today may cause a species to inhabit less than its full fundamental Grinnellian niche, violating assumptions of modeling.[7,17–19]

---

[a]Alternatively, researchers sometimes use a "pseudoabsence" sample taken from all pixels lacking a record of the species.

Table 2. Ecological niche modeling and species distribution modeling

Ecological niche modeling	Modeling of the [existing fundamental] niche of a species in environmental space with the intent of estimating the areas in geographic space holding suitable conditions for it [= the abiotically suitable area; termed the potential distribution by many workers], whether or not the species truly occupies those areas (Figs. 1 and 2). In practice (and subject to critical and clearly stated assumptions[7]), this is typically carried out by modeling the [abiotically] suitable conditions [= existing fundamental niche] for the species using: (1) records of its presence; and (2) [scenopoetic] environmental variables not affected by the presence of the species; these data sets are assembled via selection of a study region matching the assumptions of niche modeling; see Pitfall 4. The niche model is then applied to geography to identify the [abiotically] suitable area for the species. When desired, the prediction of the [abiotically] suitable area can be processed subsequently to estimate other distributional areas for the species—for example, and often of special interest, its range [= occupied distributional area]—by considering information regarding dispersal and biotic interactions. Niche models hold predictive ability across space and time.
Species distribution modeling	Modeling with the intent of estimating the range [= occupied distributional area] of a species in geographic space directly, without first passing through an estimate of its [existing fundamental] niche in environmental space or the corresponding [abiotically] suitable area in geographic space. In practice, this is typically carried out by modeling the range [= occupied distributional area] of the species using: (1) records of the presence of a species, and sometimes also information regarding sites where it is absent; and (2) various kinds of variables, including [scenopoetic] environmental variables not affected by the presence of the species, and/or spatial variables (e.g., latitude and longitude); these data sets are assembled via selection of a study region matching the assumptions of species distribution modeling and likely violating some of those of niche modeling (see Pitfall 4), resulting in models that include the effects of correlations of the two data sets with dispersal limitation and biotic interactions. The species distribution model is then applied to geography to identify the range [= occupied distributional area] of the species. The prediction of the range [= occupied distributional area] cannot be processed subsequently to estimate the [abiotically] suitable area for the species. Species distribution models are not designed to hold predictive ability across space or time.

Note: Much confusion exists in the literature regarding the terms *ecological niche modeling* and *species distribution modeling*. Although some workers do not recognize a difference between the two (generally using the term species distribution modeling for applications that are clearly niche-based), others maintain a conceptual distinction, requiring two different terms for clarity.[7] Following the latter perspective, I offer the definitions below to provide unambiguous meaning for my usage of "ecological niche modeling" of Grinnellian niches (footnote *b*). Following these definitions, many (but not all) recent studies conducting what the respective authors termed "species distribution modeling" would be recategorized as having carried out "ecological niche modeling," at least in intent. Nonspecialists may derive more clarity from reading without inclusion of the terms supplied in brackets for the specialist.

When necessary in a particular study (which is often the case), this prediction of suitable areas then can be processed to take into account dispersal barriers and the distributions of key biotic interactors (e.g., competitors, parasites, or mutualists) to estimate the species' occupied distributional area (or range, sometimes termed the realized distribution;[19,20] Figs. 1 and 2). In contrast to niche models,

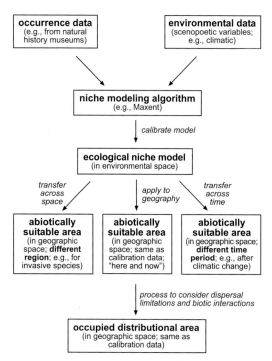

Figure 1. Diagram showing various steps involved in ecological niche modeling. Note the following sequential steps: data acquisition; calibration of the ecological niche model (in environmental space); application to geographic space (including possible transferral to other geographic regions or time periods); and processing to consider dispersal limitations and biotic interactions.[7]

approaches that estimate the species' distribution directly do not aim to predict in other places or time periods.

Most scientists implicitly or explicitly acknowledge that all models of nature remain imperfect—but that some (like, hopefully, ecological niche models) constitute quite useful approximations. Whereas the consequences of errors, or of mere inadequacies, of niche models for an academic (e.g., biogeographic) study may not incur large negative consequences for society, erroneous models for applied uses can lead to grave repercussions—for example, an endangered species poorly managed, resources for combating an invasive species misallocated, or public health policies gone awry. A chilling example comes from the related zoonotic diseases lymphatic filiariasis and loa loa in West Africa; effective prophylaxis protects against filiariasis, but its administration to a person infected with loa loa can be fatal.[21,22] Hence, niche models, applied to geog-

raphy and then processed to take into account dispersal barriers and relevant biotic interactors, must identify areas where filiariasis is transmitted but loa loa is not. As concluded recently[7] (p. 236), "such situations, in which human lives depend on being 'right,' should give pause to the prospective modeler." Because of these applications of such high importance to society, modelers and those associated with museums have the justification for making ecological niche models match the reality of nature as best as possible.

Five major pitfalls: data and methodology

So why does the full utility of ecological niche modeling remain underrealized? The answer lies in both the incomplete availability of the necessary occurrence data and the nascent nature of the field, with relatively few researchers well trained conceptually and methodologically. Over a decade ago, the field of biodiversity informatics reached a consensus regarding the plan of action for making high-quality museum data available over the internet and developed the requisite technology (e.g., via the pioneering Mammal Networked Information System, MaNIS[23]). However, despite impressive progress, implementation of that vision remains vastly incomplete.[3,24] In addition, researchers often implement the techniques poorly, usually because of lack of knowledge (and a paucity of clarity and consensus in the literature). I highlight five particular pitfalls that limit the effective use of ecological niche modeling, expanding upon each in turn and noting their interrelatedness: (1) incorrect taxonomic identifications; (2) lacking or inadequate databasing and georeferences (coordinates of latitude and longitude); (3) effects of sampling bias across geography; (4) selection of the study region; and (5) model evaluation to identify optimal model complexity (Table 3). The first three concern data quality and availability, where I draw attention to the increased necessity for (and value of) high-quality data, in light of recent progress in ecological niche modeling. In contrast, the latter two represent conceptual and methodological issues in niche modeling. Whereas other works expand on various such topics in much greater detail,[7,25] I highlight these two because of their cross-cutting nature and the strong associated repercussions when transferring a model to another place or time.[13] Other important areas, especially

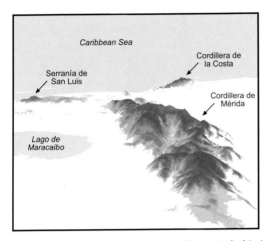

Figure 2. Example of the prediction of a species' abiotically suitable area generated by an ecological niche model. Areas shown in red indicate those estimated as suitable for the rodent *Nephelomys caracolus* in northern Venezuela based on climatic variables; the prediction appears draped over a three-dimensional representation of elevation, with increasingly dark tones indicating progressively stronger predictions for the species.[17] In addition to documented areas inhabited by *N. caracolus* in the Cordillera de la Costa, the model indicates abiotically suitable areas in the Cordillera de Mérida (which holds records of the congener *N. meridensis*) and the Serranía de San Luis (which lacks published records of either species).

improved environmental data such as remotely sensed information from satellites, offer great promise for improving the predictions of niche models,[26,27] but I present these five, which I consider fundamental to progress in the field. They apply to all niche-modeling algorithms using presence-only data along with a background (or pseudoabsence) sample of the environments available in the study region. In addition, some are germane for techniques that use presence-only occurrence data without even taking a background or pseudoabsence sample, as well as for modeling based on presence–absence and abundance data. Those concerning data availability also, of course, hold relevance for myriad other uses of museum data, far beyond those associated with ecological niche modeling.[28] Fortunately, despite the gravity of these five obstacles, all are surmountable.

Pitfall 1: Incorrect taxonomic identifications

Taxonomic knowledge—even of relatively well-studied groups such as birds, mammals, and butterflies—remains inadequate. Termed the Linnaean shortfall, this problem is especially marked in tropical regions of the greatest biodiversity.[2,29] As a result, the extremely variable quality of species identifications for museum specimens plagues their use in niche modeling.[5] For example, in my research on the spiny pocket mice (*Heteromys*) of Colombia, I found that circa one-third of the specimens lacked correct identifications, with misidentified specimens sometimes (falsely) "documenting" the species hundreds of kilometers from where it actually occurs.[30] Furthermore, whereas only two species were known from South America prior to that date, my collaborators and I subsequently described three species of South American *Heteromys* new to science based on existing (previously misidentified) specimens.[31–33] Reading of the literature and conversations with other taxonomists indicate that such underestimated and mischaracterized diversity represents the norm for small nonvolant mammals throughout the tropics. These anecdotes exemplify the fact that many studies using identifications from museum databases in reality model entities that do not represent the intended species. In a nutshell, taxonomic studies (often termed *revisionary taxonomy*) are necessary to determine what the real species are and identify each specimen correctly. Furthermore, in museum databases, each identified specimen requires associated fields indicating the source and trustworthiness of the identification. The justifications for rectifying taxonomic inadequacies of museum data sources, and the strategies for doing so, intertwine with those for the second pitfall, so I consider them together later.

Pitfall 2: Lacking or inadequate databasing and georeferences

Despite remarkable progress, the overwhelming majority of museum holdings linger undigitized, often with nonexistent or inadequate georeferences.[5,24,c] Overall, vertebrates and plants enjoy much higher rates of databasing, but even with vertebrates, information on the holdings of many critical, vast collections often remains in paper format. Furthermore, even when species are well characterized

---

[c]Although documentation of biodiversity via continued fieldwork and collection of museum specimens constitutes a critical endeavor, especially in tropical regions and for areas experiencing high rates of habitat loss,[34] I focus here on the need for making existing biodiversity information available for ecological niche modeling.

Table 3. Terms related to major pitfalls hindering ecological niche modeling

Term	Definition
Georeference	Coordinates of latitude and longitude (or another system) indicating the position of a point in space.
GPS (global positioning system)	A device that detects information from satellites to determine the current position of the user on Earth. Other features often include the ability to calculate the distances and bearings of other sites relative to the current position.
Sampling bias	Variation in the probability that a site has been sampled by biologists. Generally, such bias corresponds to accessibility in geographic space and often also leads to sampling bias in environmental space.
Transferability	The application of a model (calibrated in one region) to another place in geography and/or to another time period.
Model complexity	The level of detail of a model, here often in terms of how many variables are included, and the weights assigned to them (but also can include how complex the functions are that model a species' response to particular variables).

taxonomically, identifications are correct, and online museum databases exist, ecological niche modeling requires a high-quality georeference for each specimen (typically, geographic coordinates of latitude and longitude). This bottleneck in data availability hinders modeling at least as much as the previously mentioned inadequacies. In addition to developing technology to serve museum data on the internet, the biodiversity informatics community also established suggested standards for georeferencing.[23] First of all, georeferences should be given to the greatest precision possible. Second, a radius of possible error must be associated with each georeference, so that the user can filter records, discarding all that are not of sufficient quality for the study at hand.

I provide another anecdote from *Heteromys*, for one particularly important, yet difficult to locate, locality in Colombia: "Serranía del Darién, Alto de Barrigonal." The only record for the *H. desmarestianus* species group in South America corresponds to this site. The Serranía del Darién, a mountain range that is geologically Central American, runs along the border between Colombia and Panama, but I could not find the precise locality Alto de Barrigonal on any map or in any gazetteer. Later, I noticed that the staff of the museum housing the specimen had georeferenced it—to degrees, minutes, and seconds! Because they could not find the specific locality either, coordinates had been assigned for the middle of the mountain range. However, without an

associated possible error radius, this (otherwise reasonable) georeference was useless for most studies or, worse, perhaps even hugely misinformative. Fortunately, upon consultation, the original collector provided specific information that allowed location of the site with a fairly small radius of error.[30]

Improving the quality of georeferences holds great promise for increasing the utility of niche models.[35] Despite substantial progress with automated georeferencing,[36] determining coordinates by consulting detailed maps, field notes in museum archives, and the original collectors allows for substantially more-accurate coordinates.[33,37] Such efforts often are necessary to arrive at coordinates that are both sufficiently precise and accurate to match the approximate resolution of currently available climatic data interpolated from weather stations.[38] Other data sources, especially remotely sensed data regarding vegetation and land cover derived from satellites, may require even finer georeferences, such as those from GPS readings. Fortunately, many specimens (of various species) often come from the same locality, so not every specimen needs to be georeferenced individually.

Of course, inadequacies of taxonomic knowledge, databasing, and georeferencing (Pitfalls 1 and 2) were recognized in the past, with pioneering and visionary documents produced by the overlapping communities of taxonomic/systematic biology and biodiversity informatics, outlining the steps necessary to rectify them while taking into

Table 4. Issues related to calibrating ecological niche models

Term	Definition
Spatial bias	Bias (see sampling bias; Table 3) in geographic space.
Environmental bias	Bias (see sampling bias; Table 3) in environmental space (e.g., for the environmental variables used in ecological niche modeling).
Noise	Random variation due to the effects of sampling, without any systematic bias.
Signal	The true information concerning an entity being studied. Here, from data representing a species' niche accurately.
Calibration	The step or steps involved in forming a model, here one that estimates a species' niche based on occurrence data and environmental variables.
Evaluation	Here, the use of data not used in model calibration in order to determine model performance and significance, ideally using evaluation data fully independent of those employed in model calibration.
Sampling effort	Generally, the strength or intensity of sampling by biologists, but also may include consideration of the suite of techniques employed.
Target group	Those species that can be observed or collected with the same sampling techniques as the focal species of interest.

account necessary safeguards regarding the online dissemination of sensitive information (e.g., for endangered species vulnerable due to the pet trade[3,39]). Indeed, some museums and taxonomic communities have achieved spectacular success in databasing, and sometimes also georeferencing. However, progress for other museums and groups of organisms lags behind, especially with regard to the level and quality of georeferencing.[24] Furthermore, almost all groups of organisms still suffer from the Linnaean shortfall (with the lack of trained and employed systematists hindering progress), especially for tropical regions and for hyper-diverse invertebrates.[2]

The necessary philosophies and technologies exist for all of these challenges, awaiting implementation once advocacy leads to the necessary funding. The excellent progress in databasing and georeferencing should continue, and advancement in taxonomic studies must accelerate. Immense investment in specimen-based taxonomic studies and in pervasive digitization and georeferencing of museum holdings should lead to high-quality occurrence data available over the internet for a wide array of plants and animals.[d] In my opinion, such

funding can be justified—by making the value of the resulting data *to society as a whole* understood broadly by policy makers and governmental officials, especially those making budgetary decisions. The literature from the past few years makes it increasingly clear that those museums and taxonomic groups that achieve high-quality data available online will get their data used (showing the relevance of museum collections to society), whereas those that do not, will not. Furthermore, the better the data, the more realistic the resulting models—providing additional incentive for improvement in data quality for museum-based scientists concerned with the misuse of inadequate data if served online. The critical applications of museum data via ecological niche modeling represent an opportunity for museums to contribute information and solutions to key societal issues, as well as a compelling justification for investment in the taxonomic studies and concomitant discovery of biodiversity that have been and remain a core mission of museums.

Pitfall 3: Effects of sampling bias across geography

The very nature of the collecting expeditions that have led to the unique and irreplaceable museum holdings that constitute our primary documentation of biodiversity creates a major problem for ecological niche modeling (Table 4)—one that

---

[d]I leave to others better capable the calculations regarding the monetary amount of funding necessary.

researchers must, and can, address in order to produce realistic models.[1,5] Sampling effort varies tremendously across the globe, for example, with much higher effort in populated areas, temperate regions, and areas near roads and rivers that serve as access points.[40,41] This spatial bias affects the model-calibration process greatly (and also leads to problems for evaluating model quality, see later). The most serious problems occur when biases in the environments sampled accompany the spatial biases. The aim, of course, is to model the requirements of the species (its niche), not any bias in collection effort (in space and/or environment). When sampling effort across geography can be quantified, it should be integrated into the model-calibration process itself. Although the information necessary for direct quantification of sampling effort (e.g., via consultation of detailed field notes) does not exist for most museums, one suitable proxy for it does.

Because museum specimens themselves represent the product of sampling, they collectively can serve as a surrogate of sampling effort. This requires selection of a "target group" of species that are sampled with the same techniques as the focal species of interest (e.g., small nonvolant mammals, such as rodents, marsupials, and shrews in the Neotropics[42]). Such information—which indicates which areas, and hence environments, have been more thoroughly sampled—can then be integrated into model calibration, in effect correcting for sampling bias.[43,e] The beauty lays in the fact that digitization and georeferencing of all records of such a group (rectification of Pitfall 2) allows researchers to circumvent the serious problems posed by Pitfall 3, further leveraging such efforts![f] Fortunately, for this use, the specimen identifications only need to be correct for the focal species being modeled, with specimens of other species simply identified as members of the target group. Hence, this strategy can be implemented before taxonomists fully rectify the Linnaean shortfall (Pitfall 1).

Pitfall 4: Selection of the study region
Determining the relevant study region for model calibration represents a topic of great importance— especially when transferring a niche model to another place or time. For years, this puzzling question did not seem a primary determinant of model

quality. However, with the advent of advanced techniques capable of fitting very complex models, it came to the forefront. Recent research has clarified that environmental data from regions that may hold suitable conditions but in which the species is absent for other reasons should not be included in background samples (which are intended to represent the environments available to the species).[17,44] Specifically, a species may be absent from such areas due to dispersal barriers or because biotic interactions lead to a negative population growth rate for the focal species (for example, because of the presence of a competitor or the lack of a key mutualist[18]). Environmental information from such areas provides false negative signal that thwarts efforts to model

---

[e]Lacking data regarding sampling effort (quantified directly, or approximated via indices calculated from data regarding the target group), modelers face a quandary: to reduce bias (geographic and possibly environmental) but not signal (correct information regarding the species' niche). To ameliorate problems associated with biased sampling effort, researchers can filter localities spatially, for example, removing nearby localities (e.g., maintaining the largest set of localities possible, subject to the constraint that all lie at least x km from each other[17]). Determination of the appropriate distance x remains arbitrary at present and likely depends on the level of sampling bias and the heterogeneity of the environment (which itself likely varies across the study region).[7] If the chosen x is too small, not enough bias will be removed; conversely, if it is too large, the species' true niche signal will be diluted. Hence, although spatial filtering represents a helpful approach when no other solution is available, inclusion of information regarding sampling effort in the model-calibration process (e.g., via information on target group) constitutes a far-superior tactic.[42,43]

[f]Information regarding the target group provides an additional valuable benefit for ecological niche models: the possibility of testing for artifactual absences. The projection of a niche model onto geography indicates abiotically suitable areas for the species, but dispersal limitations[19] or biotic interactions[20] may limit it from occupying all abiotically suitable areas. In order to identify such cases, researchers must consider whether sampling effort has been sufficient to demonstrate a species' absence from a region that holds abiotically suitable conditions but lacks records of the species. Such tests for artifactual absences can be accomplished using indices of sampling effort provided by data regarding the target group.[42]

Table 5. Challenges related to transferring an ecological niche model to another place or time period and to model evaluation

Term	Definition
Interpolation	Here, prediction between known values of an environmental variable.
Extrapolation	Here, prediction into environmental values beyond the range (in environmental space) of the geographic area on which the model was calibrated (common when a model is applied to cross-time or cross-space situations).
Nonanalog environments	Environmental conditions (often climatic) in one place or time period that do not exist in another place or time (e.g., that used in model calibration). In ecological niche modeling, non analog environments require extrapolation in environmental space to make a prediction, which generally should be interpreted with great caution.
Truncated response curves	Curves of the species' response to a particular variable that do not include the full domain of an environmental variable. Such a situation can lead to nonanalog environments when transferring a model to another place or time period.
Performance	Characterization of how well or poorly a model predicts independent data, but not necessarily including statistical assessment of model significance.
Significance	Determination via statistical tests whether or not predictions of evaluation data differ from a random null hypothesis with a particular level of probabilistic confidence. Often based on some measure of model performance, tests of model significance typically assess whether the model predicts evaluation data better than random expectations (one-tailed hypothesis).
Overfitting	The situation when model complexity is excessive and a model shows close fit to calibration data but is less able to predict independent evaluation data. Note that overfitting can be to noise and/or to sampling bias.

the species' niche and the corresponding abiotically suitable areas in geography.[g]

Critically, these smaller study regions necessary for model calibration in the "here and now" highlight a key limitation of applying niche models to other places or time periods: the problem of extrapolating a model (in environmental space) to make predictions for environmental conditions that do not exist in the calibration study region (e.g., nonanalog climates; Table 5[13,45]). For example, a species may survive and reproduce very well at the

warmest temperatures present in the study region, but how can we estimate what its response would be to even warmer temperatures on another continent or under future climatic change? This problem, often termed "truncated response curves," has no easy solution, with the required laboratory or greenhouse physiological experiments laborious and seldom feasible.[46] Hence, in these cases, the researcher must make some assumption regarding the species' response in order to derive a prediction from the model.[17] In practice, some algorithms flag affected pixels of a cross-space or cross-time prediction, indicating where extrapolation in environmental space occurred and how strong an effect it had on the prediction—and therefore, where caution should be taken when interpreting predictions.[47] Although few current studies take into account these paramount principles of study-region selection and extrapolation in environmental space, they now exist clearly in the literature, and I predict that their consideration will become standard. They also

[g]In most cases, it will be difficult to identify such regions operationally in great detail (in order to exclude them from the study region used for model calibration). However, fairly reasonable study regions can be approximated (for example, by taking into account the distributions of major vegetation types), matching the relevant assumptions of modeling much more closely than the excessively expansive study regions often used presently.[17]

embody key principles for model evaluation,[7] as shown later.

Pitfall 5: Model evaluation to identify optimal model complexity

Researchers should demonstrate good performance (and statistical significance) for niche models before interpreting and using them for academic or applied uses, especially for those models that require transferral to another place or time (Table 5). Many factors can lead to poorly performing models, but the need to achieve (or at least approximate) optimal model complexity cuts across all algorithms and uses and is especially germane for cross-space or cross-time studies.[7] Unfortunately, evaluations that allow valid quantification of model performance and identification of optimal model complexity often evade researchers.[h] Having only presence-only data and a background sample (i.e., no information regarding sites where the species is absent)—plus the very objective of niche modeling: identifying the abiotically suitable areas for the species, rather than its occupied distributional area—seriously limits the options for model evaluation. However, several relevant strategies exist.

The overreaching principle of model evaluation should be to determine whether the model predicts independent data well and—for many applications—whether it has the ability to predict across space and/or time. A good model describes the species' requirements sufficiently (and better than a random prediction) but does not overfit to the peculiarities—be they bias or noise—of the calibration data. Hence, with techniques capable of producing complex models, controlling model complexity—avoiding overfitting—becomes vital.[6,7] This is especially important given the pervasive biases of biodiversity data and the small sample sizes available for many species, the latter of which

leads to concomitant problems regarding noise. Algorithmic settings that influence model complexity (e.g., the level of regularization in Maxent[50]) can affect model output greatly, creating the need to vary model parameters and select those that lead to the highest performance in evaluations made with independent data.[51–53]

Meaningful evaluations with these goals assess a model's prediction using localities that are independent of those used to calibrate it.[i] For example, researchers should not use evaluation localities that lie close to the calibration ones. That nonindependence inflates estimates of performance and significance because such sites represent the same environmental information without constituting truly independent data points.[49,j] In an even broader sense, transferring a model to another place or time period (e.g., after climatic change) requires knowledge that the model performs well under some kind of transferral. Occurrence data from other time periods generally do not exist for studies of the effects of climatic change. In such cases, the only option for evaluating transferral with independent data is to transfer across space, for example, via spatially structured subsampling of available occurrence records into

---

[h]Confusion regarding evaluations of model quality permeates much of the literature, primarily because researchers frequently use measures appropriate for presence–absence evaluation data but that misrepresent reality when applied to presence only evaluation data (along with a background or pseudoabsence sample). Perhaps most importantly, false positive rates (= commission error rates) suffer from overinflation, often very substantially.[48] Myriad critical technical points regarding model evaluation appear elsewhere.[7,25,49]

[i]As intimated above (Pitfall 4), theory indicates that the study region appropriate for model calibration also constitutes that appropriate for model evaluation.[7] Violation of this principle pervades the literature, leading to a double whammy: poor models (because of an improper calibration region) that appear to be very good (because of the same, inappropriate, region used for evaluation). Typically, this occurs with extremely large study regions that include extensive areas of abiotically suitable conditions from which the species is absent due to dispersal limitations associated with contingent events of its evolutionary history[19] (although the limitations also could derive from biotic interactions that lead to a negative population growth rate for the focal species[20]). The resulting false negative signal from such regions leads to models that vastly underestimate the abiotically suitable area for the species; concomitantly, the evaluation erroneously suggests that such a prediction is correct, because no occurrence record for the species exists from those regions (which lie on the other side of the dispersal barrier). [j]Spatial filtering of occurrence data (mentioned above for Pitfall 3) likely ameliorates this problem, but the degree of filtering remains arbitrary.

calibration and evaluation datasets.[6,7,54,k] Although few recent studies employ evaluation strategies effective in identifying optimal model complexity and the ability to predict across space and/or time, I predict that they will become pervasive.

Agenda: Making museum data and ecological niche modeling ready to address critical environmental issues of the 21st century

Despite these largely cautionary reflections, tremendous opportunity for progress exists, both with regard to data availability and the implementation of ecological niche modeling. Concerning data availability, museum holdings only will be maximally useful to society when high-quality information on those specimens is available across regions and taxonomic groups, ready to be accessed *when the particular problem for a species of interest presents itself*. For example, studies of an emerging zoonotic disease or recently detected invasive species call for quick action. In such a situation, society needs to be able to count on the data being ready, rather than requiring a special (and often prohibitively slow) effort to clarify the taxonomy and digitize and georeference the records of the species of interest (Pitfalls 1 and 2). Furthermore, correcting for collection biases (a critical issue in niche modeling; Pitfall 3) will be possible when data for the whole target group exist for all the museums providing records of the focal species itself.

In my view, museums (or better yet, consortia of museums) can make effective arguments for major funding to realize these societal needs. The key lies in leveraging unique museum holdings to justify investment in taxonomic studies, identification of specimens, databasing, and georeferencing of localities. Particular subgroups more likely to be of economic or medical relevance (e.g., rodents or flies) and those already in a relatively advanced state of taxonomic knowledge surely can sell these ideas more easily. However, in principle, the utility of investments in making museum data useful to society should ring clear across taxonomic groups and geographic regions. For the short term, I suggest continued large-scale, across-the-board databasing and georeferencing initiatives, with a variety of high-priority taxonomic projects realized in parallel. Such progress will allow niche modeling for the most-critical focal species, taking advantage of digitized and georeferenced records of the full target group in order to correct for collection biases (even if the taxonomic information of some members of the target group remains incomplete). In the medium term, further taxonomic efforts should fill in the gaps in high-quality identifications as rapidly as possible. Although progress rectifying the Linnaean shortfall surely will proceed unevenly with respect to taxonomic group and geographic region, effective advocacy should lead to ever-broadening coverage over time.

Parallel to such data-availability efforts, society needs advances in ecological niche modeling, with regard both to technology and—especially—to human capacity. For the former, the field requires software that achieves an appropriate balance between automation and supervision. Currently, many software packages implementing niche-modeling algorithms provide default settings that allow efficient processing of data automatically, yielding niche models for many species simultaneously. However, such ease of use also promotes ease of misuse. Biologists with knowledge of the species and geographic area must select the relevant study region for model calibration carefully (Pitfall 4). Similarly, well-trained modelers must supervise the evaluation process, because default settings do not necessarily lead to models with optimal levels of model complexity (Pitfall 5). At present, it remains time consuming to implement the steps of model calibration and evaluation necessary to produce and document highly performing models, even for skilled users programming batch files via code. This situation calls for software that automates repetitive aspects of the process, while allowing (and forcing) the user to provide input when critical biological and conceptual decisions need to be made. Ideally, such software will be general with respect to the actual modeling algorithm or algorithms used to create the model—that is, umbrella tools that can be employed in concert with any particular modeling algorithm desired.

---

[k]As is appropriate for tests of transferability, these evaluations require the same caveats and assumptions related to truncated response curves and nonanalog environments mentioned earlier,[45] as well as the assumption of no differences in inherited niche physiology across geography or time.[7]

Complementarily, the field must produce a much larger number of scientists capable of building and applying high-quality niche models, as well as a broad community able to appraise their quality and utility. Rather similar to the trend in job searches in museums and universities in the 1980s and 1990s, which often targeted taxonomists and evolutionary biologists with molecular skills (e.g., protein electrophoresis and later DNA sequencing), I predict that the next decade will see a wave of openings specifically desiring scientists fluent in ecological niche modeling and associated spatial analyses.[11] To train this generation of researchers and expand the knowledge base of those educated before them requires both clear literature and extensive educational opportunities. Although this research area remains in extremely rapid development, the literature published over the past few years seems to be leading towards a synthesis of ideas and a maturation of methodologies.[7] Obviously, editors, reviewers, and the pool of scientists worldwide all play vital roles in helping achieve these goals via productive debate, insistence upon rigor, and a fostering of creativity. Furthermore, continued proliferation of workshops and graduate courses will be necessary to train the scientists needed by society. This training must include both the theoretical and the methodological aspects of niche modeling. In my experience, such venues bring together fascinating mixes of people excited to apply emerging technology, consider new ideas, and answer biogeographic questions regarding the systems they study.

Indeed, I have found that ecological niche modeling and biodiversity informatics attract scientists with exceptionally diverse backgrounds and interests. That, together with their intelligence and creativity, epitomizes the interdisciplinary nature of the field and has made it a fascinating area of research over the past decade. With a meeting of societal needs, stores of critical data, blossoming technologies, and capable minds, the next years provide the opportunity to realize the potential that museums and ecological niche modeling offer society. Given the enormous relevance of biodiversity and the environment to humankind over the next century—and the irreversibility of many of the biological changes that may occur—a significant component of history will depend on our success.

Acknowledgments

This work was supported by the U.S. National Science Foundation (NSF DEB-0717357 and DEB-1119915) and the Professional Staff Congress of the City University of New York (Grant 64215-00-42). I salute a long list of colleagues who have made my explorations in ecological niche modeling of species distributions both fascinating and productive. Animated conversations and debates with Miguel B. Araújo, Santiago F. Burneo, Miroslav Dudík, Jane Elith, Marcela Gómez-Laverde, Israel Gonzalez, Jr., Catherine H. Graham, Eliécer E. Gutiérrez, Sharon A. Jansa, Kenneth H. Kozak, Daniel Lew, Enrique Martínez-Meyer, Miguel Nakamura, A. Townsend Peterson, Víctor Sánchez-Cordero, Norman A. Slade, Jorge Soberón, Pascual J. Soriano, Richard G. Pearson, Steven J. Phillips, Aleksandar Radosavljevic, Christopher J. Raxworthy, Ali Raza, Robert E. Schapire, Mariya Shcheglovitova, Mariano Soley-G., and Eleanor J. Sterling led to many epiphanies, some of which took years to become clear to me. In particular, my thinking derives from immersion in the stimulating intellectual environments at the University of Kansas, the American Museum of Natural History, and the City University of New York. Final framing of these ideas occurred as a sabbatical visitor at the Museo Nacional de Ciencias Naturales in Madrid. Robert A. Boria, Douglas Braaten, Eliécer E. Gutiérrez, Aleksandar Radosavljevic, Mariano Soley-G., and an anonymous reviewer provided helpful comments on previous drafts of the manuscript. Finally, I thank my fiancé Rick for his humor and support during the past eight years of this journey and my parents and grandparents for instilling in me a fascination for the world outside and nurturing the curiosity to discover how it works.

Conflicts of interest

The author declares no conflicts of interest.

References

1. Funk, V.A. & K.S. Richardson. 2002. Systematic data in biodiversity studies: use it or lose it. *Syst. Biol.* **51:** 303–316.
2. Wilson, E.O. 2003. The encyclopedia of life. *Trends Ecol. Evol.* **18:** 77–80.
3. Soberón, J. & A.T. Peterson. 2004. Biodiversity informatics: managing and applying primary biodiversity data. *Phil. Trans. R. Soc. Lond. B.* **359:** 689–698.

4. Boakes, E.H., P.J.K. McGowan, R.A. Fuller, *et al.* 2010. Distorted views of biodiversity: spatial and temporal bias in species occurrence data. *PLOS Biology.* **8:** 1–11.

5. Newbold, T. 2010. Applications and limitations of museum data for conservation and ecology, with particular attention to species distribution models. *Progr. Phys. Geogr.* **34:** 3–22.

6. Jiménez-Valverde, A., A.T. Peterson, J. Soberón, *et al.* 2011. Use of niche models in invasive species risk assessments. *Biol. Invasions.* **13:** 2785–2797.

7. Peterson, A.T., J. Soberón, R.G. Pearson, *et al.* 2011. *Ecological Niches and Geographic Distributions.* Monographs in Population Biology 49, Princeton University Press. Princeton, NJ.

8. Zimmermann, N.E., T.C. Edwards, Jr., C.H. Graham, *et al.* 2010. New trends in species distribution modelling. *Ecography* **33:** 985–989.

9. Elith, J., C.H. Graham, R.P. Anderson, *et al.* 2006. Novel methods improve prediction of species' distributions from occurrence data. *Ecography* **29:** 129–151.

10. Kozak, K.H., C.H. Graham & J.J. Wiens. 2008. Integrating GIS-based environmental data into evolutionary biology. *Trends Ecol. Evol.* **23:** 141–148.

11. Swenson, N.G. 2008. The past and future influence of geographic information systems on hybrid zone, phylogeographic and speciation research. *J. Evol. Biol.* **21:** 421–434.

12. Thomas, C.D., A. Cameron, R.E. Green, *et al.* 2004. Extinction risk from climate change. *Nature* **427:** 145–148.

13. Elith, J. & J.R. Leathwick. 2009. Species distribution models: ecological explanation and prediction across space and time. *Annu. Rev. Ecol. Evol. Syst.* **440:** 677–697.

14. Chase, J.M. & M.A. Leibold. 2003. *Ecological Niches: Linking Classical and Contemporary Approaches.* University of Chicago Press. Chicago.

15. Holt, R.D. 2009. Bringing the Hutchinsonian niche into the 21st Century: ecological and evolutionary perspectives. *Proc. Nat. Acad. Sci. U.S.A.* **106:** 19659–19665.

16. Soberón, J.M. 2010. Niche and area of distribution modeling: a population ecology perspective. *Ecography* **33:** 159–167.

17. Anderson, R.P. & A. Raza. 2010. The effect of the extent of the study region on GIS models of species geographic distributions and estimates of niche evolution: preliminary tests with montane rodents (genus *Nephelomys*) in Venezuela. *J. Biogeogr.* **37:** 1378–1393.

18. Gaston, K.J. 2003. *The Structure and Dynamics of Geographic Ranges.* Oxford University Press. Oxford.

19. Anderson, R.P., A.T. Peterson & M. Gómez-Laverde. 2002. Using niche-based GIS modeling to test geographic predictions of competitive exclusion and competitive release in South American pocket mice. *Oikos* **98:** 3–16.

20. Anderson, R.P., M. Gómez-Laverde & A.T. Peterson. 2002. Geographical distributions of spiny pocket mice in South America: insights from predictive models. *Global Ecol. Biogeogr.* **11:** 131–141.

21. Thomson, MC.,V. Obsomer, M. Dunne, *et al.* 2000. Satellite mapping of loa loa prevalence in relation to ivermectin use in west and central Africa. *Lancet* **356:** 1077–1078.

22. Gyapong, J.O., D. Kyelem, I. Kleinschmidt, *et al.* 2002. The use of spatial analysis in mapping the distribution of ban- croftian filariasis in four West African countries. *Annals. Trop. Med. Paras.* **96:** 695–705.

23. Stein, B.R., & J. Wieczorek. 2004. Mammals of the world: MaNIS as an example of data integration in a distributed network environment. *Biodiv. Informatics* **1:** 14–22.

24. Ariño, A.H. 2010. Approaches to estimating the universe of natural history collections data. *Biodiv. Informatics.* **7:** 81–92.

25. Araújo, M.B. & A. Guisan. 2006. Five (or so) challenges for species distribution modelling. *J. Biogeogr.* **33:** 1677–1688.

26. Bradley, B.A. & E. Fleishman. 2008. Can remote sensing of land cover improve species distribution modeling? *J. Biogeogr.* **35:** 1158–1159.

27. Austin, M.P. & K.P. Van Neil. 2010. Improving species distribution models for climate change studies: variable selection and scale. *J. Biogeogr.* **38:** 1–8.

28. Graham, C.H., S. Ferrier, F. Huettman [*sic*], *et al.* 2004. New developments in museum-based informatics and application in biodiversity analysis. *Trends Ecol. Evol.* **19:** 497–503.

29. Whittaker, R.J., M.B. Araújo, P. Jepson, *et al.* 2005. Conservation biogeography: assessment and prospect. *Diversity Distrib.* **11:** 3–23.

30. Anderson, R.P. 1999 [2000]. Preliminary review of the systematics and biogeography of the spiny pocket mice (*Heteromys*) of Colombia. *Rev. Acad. Colomb. Cienc. Exactas Físicas Nat.* **23**(suplemento especial): 613–630.

31. Anderson, R.P. & P. Jarrín-V. 2002. A new species of spiny pocket mouse (Heteromyidae: *Heteromys*) endemic to western Ecuador. *Amer. Mus. Novitates* **3382:** 1–26.

32. Anderson, R.P. 2003. Taxonomy, distribution, and natural history of the genus *Heteromys* (Rodentia: Heteromyidae) in western Venezuela, with the description of a dwarf species from the Península de Paraguaná. *Amer. Mus. Novitates* **3396:** 1–43.

33. Anderson, R.P. & E.E. Gutiérrez. 2009. Taxonomy, distribution, and natural history of the genus *Heteromys* (Rodentia: Heteromyidae) in central and eastern Venezuela, with the description of a new species from the Cordillera de la Costa. *Systematic Mammalogy: Contributions in Honor of Guy G. Musser* (ed. by R.S. Voss and M.D. Carleton). *Bull. Amer. Mus. Nat. Hist.* **331:** 33–93.

34. Raven, P.H. & E.O. Wilson. 1992. A fifty-year plan for biodiversity surveys. *Science* **258:** 1099–1100.

35. Feeley, K.H. & M.R. Silman. 2010. Modelling the responses of Andean and Amazonian plant species to climate change: the effects of georeferencing errors and the importance of data filtering. *J. Biogeogr.* **37:** 733–740.

36. Guralnick, R.P., J. Wieczorek, R. Beaman, *et al.* 2006. BioGeomancer: automated georeferencing to map the world's biodiversity data. *PLoS Biol.* **4:** 1901–1909.

37. García-Milagros, E. & V.A. Funk. 2010. Improving the use of information from museum specimens: using Google Earth© to georeference Guiana Shield specimens in the US National Herbarium. *Front. Biogeogr.* **2:** 71–77.

38. Hijmans, R.J., S.E. Cameron, J.L. Parra, *et al.* 2005. Very high resolution interpolated climate surfaces for global land areas. *Int. J. Climatol.* **25:** 1965–1978.

39. Samper, C. 2004. Taxonomy and environmental policy. *Phil. Trans. R. Soc. Lond. B.* **359:** 721–728.

40. Reddy, S. & L.M. Dávalos. 2003. Geographical sampling bias and its implications for conservation priorities in Africa. *J. Biogeogr.* **30:** 1719–1727.

41. Kadmon, R., O. Farber & A. Danin. 2004. Effect of roadside bias on the accuracy of predictive maps produced by bioclimatic models. *Ecol. Appl.* **14:** 401–413.

42. Anderson, R.P. 2003. Real vs. artefactual absences in species distributions: tests for *Oryzomys albigularis* (Rodentia: Muridae) in Venezuela. *J. Biogeogr.* **30:** 591–605.

43. Phillips, S.J., M. Dudík, J. Elith, *et al.* 2009. Sample selection bias and presence-only distribution models: implications for background and pseudo-absence data. *Ecol. Appl.* **19:** 181–197.

44. Barve, N., V. Barve, A. Jiménez-Valverde, *et al.* 2011. The crucial role of the accessible area in ecological niche modeling and species distribution modeling. *Ecol. Model.* **222:** 1810–1819.

45. Williams, J.W., S.T. Jackson & J.E. Kutzbach. 2007. Projected distributions of novel and disappearing climates by 2100 AD. *Proc. Nat. Acad. Sci. U.S.A.* **104:** 5738–5742.

46. Williams, J.W. & S.T. Jackson. 2007. Novel climates, no-analog communities, and ecological surprises. *Front. Ecol. Environ.* **5:** 475–482.

47. Elith, J., S.J. Phillips, T. Hastie, *et al.* 2011. A statistical explanation of MaxEnt for ecologists. *Diversity Distrib.* **17:** 43–57.

48. Anderson, R.P., D. Lew & A.T. Peterson. 2003. Evaluating predictive models of species' distributions: criteria for selecting optimal models. *Ecol. Model.* **162:** 211–232.

49. Veloz, S.D. 2009. Spatially autocorrelated sampling falsely inflates measures of accuracy for presence-only niche models. *J. Biogeogr.* **36:** 2290–2299.

50. Phillips, S.J., R.P. Anderson & R.E. Schapire. 2006. Maximum entropy modeling of species geographic distributions. *Ecol. Model.* **190:** 231–259.

51. Elith, J., M. Kearney & S. Phillips. 2010. The art of modelling range-shifting species. *Meth. Ecol. Evol.* **1:** 330–342.

52. Anderson, R.P. & I. Gonzalez, Jr. 2011. Species-specific tuning increases robustness to sampling bias in models of species distributions: an implementation with Maxent. *Ecol. Model.* **222:** 2796–2811.

53. Warren, D.L. & S.N. Seifert. 2011. Ecological niche modeling in Maxent: the importance of model complexity and the performance of model selection criteria. *Ecol. Appl.* **21:** 335–342.

54. Araújo, M.B. & C. Rahbek. 2006. How does climate change affect biodiversity? *Science* **313:** 1396–1397.

Ann. N.Y. Acad. Sci. ISSN 0077-8923

ANNALS OF THE NEW YORK ACADEMY OF SCIENCES
Issue: *Blavatnik Awards for Young Scientists*

Protein function and allostery: a dynamic relationship

Charalampos G. Kalodimos

Department of Chemistry and Chemical Biology, Rutgers University, Piscataway, New Jersey

Address for correspondence: Charalampos G. Kalodimos, Department of Chemistry and Chemical Biology, Rutgers University, 599 Taylor Rd., Piscataway, NJ 08854. babis@rutgers.edu

Allostery is a fundamental process by which distant sites within a protein system sense each other. Allosteric regulation is such an efficient mechanism that it is used to control protein activity in most biological processes, including signal transduction, metabolism, catalysis, and gene regulation. Over recent years, our view and understanding of the fundamental principles underlying allostery have been enriched and often utterly reshaped. This has been especially so for powerful techniques such as nuclear magnetic resonance spectroscopy, which offers an atomic view of the intrinsic motions of proteins. Here, I discuss recent results on the catabolite activator protein (CAP) that have drastically revised our view about how allosteric interactions are modulated. CAP has provided the first experimentally identified system showing that (i) allostery can be mediated through changes in protein motions, in the absence of changes in the mean structure of the protein, and (ii) favorable changes in protein motions may activate allosteric proteins that are otherwise structurally inactive.

Keywords: protein allostery; protein dynamics; NMR spectroscopy

Introduction

A fundamental question in allostery is how perturbation at one site is transmitted through the protein to remote sites to effect binding or enzymatic activity regulation.[1–5] It is generally thought that changes in protein shape and bonding interactions, which are considered to contribute primarily to enthalpy, are necessary to propagate binding signals to remote sites.[1] This purely mechanical view of allostery—invoking only structural changes—was advanced and established as the classical view of the phenomenon by the early crystallographic work on allosteric systems, such as hemoglobin and several enzymes.[1] However, because allostery is fundamentally thermodynamic in nature, long-range communication may be mediated not only by changes in the mean conformation (enthalpic contribution), but also by changes in the dynamic fluctuations about the mean conformation (entropic contribution).[6] Indeed, the possibility of allosteric regulation through dynamic (entropic) mechanisms has long been recognized,[7] at least at the theoretical level, but has been difficult to prove experimentally.[8,9]

Nuclear magnetic resonance (NMR) spectroscopy is one of the most powerful tools for the characterization of biomolecular systems. A unique aspect of NMR is its capacity to provide integrated insight into both the structure and intrinsic dynamics of biomolecules.[10] In addition, NMR can provide site-resolved information about the conformation entropy of binding,[11–13] and about energetically excited conformational states.[14,15] NMR characterization of CAP has recently provided an unprecedented insight into the intimate link between protein intrinsic motions and function.[16–18]

CAP holds an esteemed role in biochemistry history.[19] It has been described in countless textbooks as a canonical example of effector-mediated allosteric regulation as well as a prototypic activator of transcription initiation.[20] CAP is a 50 kDa homodimer with each subunit organized in two distinct domains: (i) an N-terminal cAMP-binding domain (CBD) (residues 1–136), which contains the cyclic nucleotide–binding module and a long α-helix (C-helix) that mediates dimerization through formation of an intersubunit coiled coil, and (ii) a C-terminal DNA-binding domain (DBD) (residues 139–209), which contains a helix-turn-helix (HTH)

doi: 10.1111/j.1749-6632.2011.06319.x

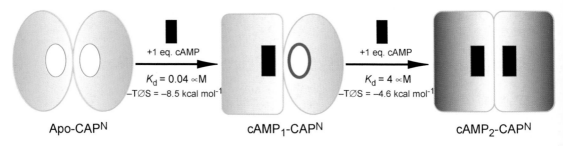

Figure 1. Effect of the sequential, anticooperative binding of cAMP to CBD of CAP (CAP^N). Binding of the first cAMP changes the structure of the liganded subunit but has no effect on the mean structure of the unliganded subunit. In contrast, the slow motions (μs-ms) of the unliganded subunit are enhanced (thicker red line). Binding of the second cAMP suppresses the fast motions (ps-ns) on both subunits (blue color). As a result, binding of the second cAMP incurs an unfavorable conformational entropy change, which reduces the affinity of the second cAMP. Thus, the negative cooperativity is entirely of an entropic nature (from Ref. 16).

motif for binding to DNA.[21] The two domains are linked by a hinge region (residues 137–138). cAMP elicits an allosteric transition that switches CAP from the "off" state, which binds DNA weakly and nonspecifically, to the "on" state, which binds DNA strongly and specifically.[17,22]

Dynamically driven protein allostery

Two cAMP molecules bind to dimeric CAP with negative cooperativity.[16] We exploited the strong negative cooperativity of cAMP binding to CBD of CAP to "freeze" binding conformations at intermediate stages.[16] The intermediate stages are the key conformational states, as they "contain" the information about how the allosteric sites communicate.[23,24] These intermediate states are typically difficult to stabilize and characterize. Based on chemical shift perturbation, which is a very sensitive measure of changes in the average protein conformation, we found that binding of the first cAMP to CAP did not induce long-range structural effects to the unliganded subunit (Fig. 1). Thus, the mean conformation of the unfilled cAMP site at the unliganded subunit of the intermediate-state complex is not at all affected by the presence of the first cAMP, suggesting that the contribution of the "structural" component to the observed negative cooperativity is negligible.

To understand how protein dynamics adjust along the allosteric reaction coordinate, the backbone motions of CAP were measured as a function of the cAMP ligation state over a wide range of functionally relevant timescales by measuring relaxation rates by NMR. Slow domain motions on the micro- to millisecond (μs–ms) scale are bio-

logically very important because they are close to the timescales on which functional processes take place, and they indicate the presence of energetically excited conformational states.[25–27] Motions on the pico- to nanosecond (ps–ns) fast timescale are also important because of their strong effect on the entropy of the system.[28,29] In contrast to the absence of structural changes, the intrinsic motions of CAP residues in the unliganded subunit were strongly affected upon cAMP binding (Fig. 1). Thus, it appears that the unliganded subunit "senses" the presence of the ligand (cAMP) in the liganded subunit only through changes in protein motions but not through structural changes. Notably, the data indicated stimulation of fluctuations about the mean structure on the slow (μs–ms) timescale, despite the fact that no change in the mean structure was detected. It is particularly noteworthy that slow and fast motions of residues located at distant regions were affected in the absence of a visible connectivity pathway. This result further undermines the mechanical view of allostery, wherein binding effects are assumed to propagate through a series of conformational distortions. Rather, the ligand-induced redistribution of the protein's dynamic fluctuations affects regions linked by cooperative interactions, thereby providing a means of propagating the allosteric signal to the distal site even in the absence of structural changes.

Interestingly, the thermodynamic basis of the observed anti-cooperative binding of cAMP to CAP is entirely of entropic nature. This finding suggested that the observed extensive changes in protein motions upon sequential cAMP binding were the most probable source of the large difference in entropy change between the two cAMP binding steps.

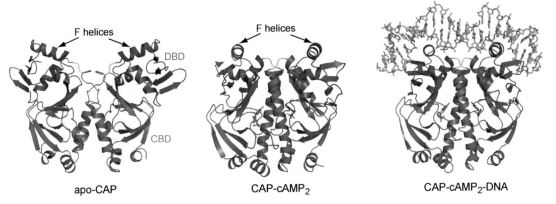

Figure 2. Structural basis for the cAMP-induced allosteric activation of CAP. The structures of apo-CAP, CAP–cAMP$_2$, and CAP–cAMP$_2$–DNA are shown. cAMP-free CAP does not bind to DNA because the orientation of the F helices is out of register with respect to the major grooves of the DNA. cAMP binding elicits an allosteric transition to the DBDs, which undergo a rigid-body rotation that place the F helices in the correct position to interact with successive DNA major grooves.

Estimation of the conformational entropy from changes of order parameters[30,31] indeed confirmed that the calorimetrically measured difference in entropy between the two sequential binding steps was primarily due to alterations in protein flexibility. The larger conformational entropic penalty in the second step greatly decreases the total entropy of the system, resulting in weaker and, thus, anti-cooperative binding.

The anti-cooperative binding of cAMP to CAP cannot be accounted for on the basis of structural changes, but can be satisfactorily explained by the experimentally determined changes in flexibility and, thus, of binding entropy. This could very well be the case for other systems showing negative cooperativity for which crystallographic studies have not revealed any obvious structural effect on the unfilled subunit in the intermediate complex.[16]

Dynamic activation of protein function

A central postulate of the classical allostery theory is that effectors control protein activity by inducing and stabilizing specific conformational states with distinct binding or enzymatic activity. Failure of an effector to induce and stabilize the active conformation is expected to result in suppressed protein activity. Thus, allosteric regulation is thought to be mediated exclusively through structural transitions that select and stabilize appropriate conformational states.

Structural characterization of CAP in the different ligation states indicated that cAMP-elicited acti-

vation of CAP for DNA binding requires the DBDs to undergo a pronounced conformational change by undergoing a ~60° rigid-body rotation (Figs. 2 and 3A).[17,22] Noneffector molecules, such as cGMP, bind to DBD but fail to elicit the active DBD conformation and, as a result, CAP is not activated by cGMP binding (Fig. 3B).[18] Clearly, any CAP variant adopting the inactive, DNA-binding incompetent conformation is expected not to bind to DNA.

Interestingly, we discovered a CAP mutant (S62F) that throws this general view in question. cAMP binding to the CAP mutant fails to elicit the allosteric transition and DBD remains in the inactive conformation (Fig. 3C).[18] Apparently, the S62F substitution decouples the long-range structural communication between CBD and DBD, and as a result the relative orientation of DBD is not allosterically affected by cAMP binding. Surprisingly, CAP-S62F–cAMP$_2$ binds to DNA with very strong affinity, despite the fact that the DBD adopts the DNA-binding incompetent conformation. Why does this structurally inactive protein bind to DNA?

In fact, results from relaxation dispersion experiments suggest that the active conformation is present in CAP-S62F–cAMP$_2$, but it is marginally stabilized, and, thus, poorly populated (~2%) (Fig. 3C). Even in this case, CAP-S62F–cAMP$_2$ would be expected to bind to DNA with at least 50-fold lower affinity than wild-type (WT) CAP–cAMP$_2$.[32] Nevertheless, CAP-S62F–cAMP$_2$ binds to DNA very tightly, with complex formation being entirely driven by a very favorable conformational

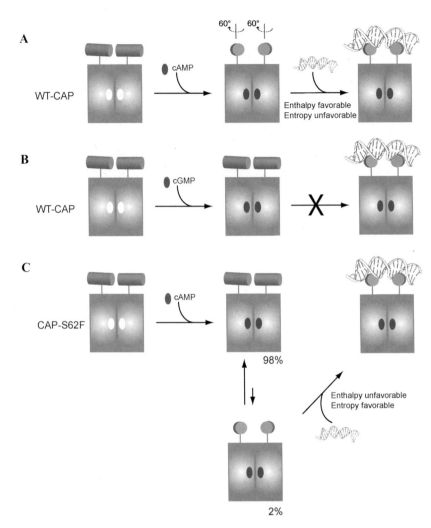

Figure 3. Reaction pathways for cAMP-mediated CAP activation and DNA binding. (A) cAMP binding to WT CAP elicits the active conformation so that DBD becomes structurally poised to interact favorably with DNA. Complex formation is strongly enthalpically favored and entropically unfavorable. (B) cGMP binds strongly to CBD in CAP but fails to elicit the allosteric transition in the DBDs. (C) cAMP binding to CAP-S62F stabilizes only marginally the active conformation, which is poorly populated ($\sim$2%). Because DNA binds to the active conformation of CAP with many orders of magnitude stronger affinity than to the inactive conformation, DNA will bind selectively to the active, low-populated DBD state and shift the population from the inactive DBD to the active DBD conformation. DNA binding to CAP-S62F–cAMP is entirely driven by entropy, which is dominated by favorable conformational entropy change upon DNA binding (from Ref. 18).

entropy change as measured by ITC. In sharp contrast, WT CAP–cAMP$_2$ binding to DNA is entirely driven by a large enthalpy change as it incurs a very unfavorable entropy change.

To understand the origin of this large favorable change in entropy, we sought to determine the role of dynamics in the binding process by measuring changes in order parameters for DNA binding to CAP. The data showed that DNA binding to either WT or mutant CAP proteins results in a dramatic redistribution of motions, with WT CAP–cAMP$_2$ becoming overall more rigid upon DNA binding but with CAP-S62F–cAMP$_2$ becoming much more flexible.

It is of particular interest that the NMR-measured per-residue conformational entropy data for the interaction of WT CAP–cAMP$_2$ and CAP-S62F–cAMP$_2$ proteins with DNA shows that, whereas the wild type complex is accompanied by a large unfavorable conformational entropy, formation of the mutant complex incurs a very favorable conformational entropy (Fig. 3C). Taking into account that the contribution of hydration phenomena to CAP DNA binding should be very similar for both WT CAP and CAP-S62F proteins, the large entropy difference measured by ITC for the formation of the two DNA complexes can be attributed to the dramatically different conformational entropy change of binding. These collective data provide strong evidence that CAP-S62F–cAMP$_2$ is "dynamically" activated for DNA binding.

Conclusions

Allosteric interactions were long considered to proceed through a series of discrete changes in bonding interactions and well-defined structural pathways.[1] This mechanical view has ignored the strong effect that changes in intrinsic protein motions may exert on protein activity. Accumulating evidence indicates that internal dynamics can mediate long-range communication either by means of conformational entropy changes or by linking distal sites through the modulation of slower motions that are capable of tuning binding and activity by eliciting alternate conformational states.[2,6,9,16,18,33–50] Thus, it is the inextricable link and the interplay between structure and dynamics that ultimately control protein activity.

Acknowledgment

This work was supported by the National Science Foundation (MCB-1121896).

Conflicts of interest

The author declares no conflicts of interest.

References

1. Changeux, J. & S. Edelstein. 2005. Allosteric mechanisms of signal transduction. *Science* **308**: 1424–1428.
2. Smock, R. & L. Gierasch. 2009. Sending signals dynamically. *Science* **324**: 198–203.
3. Tsai, C., A. del Sol & R. Nussinov. 2008. Allostery: absence of a change in shape does not imply that allostery is not at play. *J. Mol. Biol.* **378**: 1–11.
4. Brunori, M. 2011. Allostery turns 50: is the vintage yet attractive? *Protein Sci.* **20**: 1097–1099.
5. Maksay, G. 2011. Allostery in pharmacology: thermodynamics, evolution and design. *Prog. Biophys. Mol. Biol.* **106**: 463–473.
6. Wand, A. 2001. Dynamic activation of protein function: a view emerging from NMR spectroscopy. *Nat. Struct. Biol.* **8**: 926–931.
7. Cooper, A. & D. Dryden. 1984. Allostery without conformational change. A plausible model. *Eur. Biophys. J.* **11**: 103–109.
8. Kern, D. & E. Zuiderweg. 2003. The role of dynamics in allosteric regulation. *Curr. Opin. Struct. Biol.* **13**: 748–757.
9. Tzeng, S-R. & C.G. Kalodimos. 2011. Protein dynamics and allostery: an NMR view. *Curr. Opin. Struct. Biol.* **21**: 62–67.
10. Mittermaier, A. & L.E. Kay. 2006. New tools provide new insights in NMR studies of protein dynamics. *Science* **312**: 224–228.
11. Palmer, A., J. Williams & A. McDermott. 1996. Nuclear magnetic resonance studies of biopolymer dynamics. *J. Phys. Chem.* **100**: 13293–13310.
12. Yang, D. & L. Kay. 1996. Contributions to conformational entropy arising from bond vector fluctuations measured from NMR-derived order parameters: application to protein folding. *J. Mol. Biol.* **263**: 369–382.
13. Li, Z., S. Raychaudhuri & A. Wand. 1996. Insights into the local residual entropy of proteins provided by NMR relaxation. *Protein Sci.* **5**: 2647–2650.
14. Korzhnev, D.M., T.L. Religa, W. Banachewicz, A.R. Fersht & L.E. Kay. 2010. A transient and low-populated protein-folding intermediate at atomic resolution. *Science* **329**: 1312–1316.
15. Bouvignies, G., P. Vallurupalli, D.F. Hansen, *et al.* 2011. Solution structure of a minor and transiently formed state of a T4 lysozyme mutant. *Nature* **477**: 111–114.
16. Popovych, N., S. Sun, R.H. Ebright, & C.G. Kalodimos. 2006. Dynamically driven protein allostery. *Nat. Struct. Mol. Biol.* **13**: 831–838.
17. Popovych, N., S-R. Tzeng, M. Tonelli, *et al.* 2009. Structural basis for cAMP-mediated allosteric control of the catabolite activator protein. *Proc. Natl. Acad. Sci. U.S.A.* **106**: 6927–6932.
18. Tzeng, S.-R. & C.G. Kalodimos. 2009. Dynamic activation of an allosteric regulatory protein. *Nature* **462**: 368–372.
19. Harman, J. 2001. Allosteric regulation of the cAMP receptor protein. *Biochim. Biophys. Acta* **1547**: 1–17.
20. Lawson, C., D. Swigon, K. Murakami, *et al.* 2004. Catabolite activator protein: DNA binding and transcription activation. *Curr. Opin. Struct. Biol.* **14**: 10–20.
21. Schultz, S., G. Shields & T. Steitz. 1991. Crystal structure of a CAP-DNA complex: the DNA is bent by 90 degrees. *Science* **253**: 1001–1007.
22. Passner, J.M., S.C. Schultz & T.A. Steitz. 2000. Modeling the cAMP-induced allosteric transition using the crystal structure of CAP-cAMP at 2.1 Å resolution. *J. Mol. Biol.* **304**: 847–859.
23. Koshland, D. 1996. The structural basis of negative cooperativity: receptors and enzymes. *Curr. Opin. Struct. Biol.* **6**: 757–761.
24. Stevens, S., S. Sanker, C. Kent & E. Zuiderweg. 2001. Delineation of the allosteric mechanism of a cytidylyltransferase exhibiting negative cooperativity. *Nat. Struct. Biol.* **8**: 947–952.

25. Akke, M. 2002. NMR methods for characterizing microsecond to millisecond dynamics in recognition and catalysis. *Curr. Opin. Struct. Biol.* **12:** 642–647.

26. Palmer, A. 2004. NMR characterization of the dynamics of biomacromolecules. *Chem. Rev.* **104:** 3623–3640.

27. Mittermaier, A.K. & L.E. Kay. 2009. Observing biological dynamics at atomic resolution using NMR. *Trends Biochem. Sci.* **34:** 601–611.

28. Jarymowycz, V. & M. Stone. 2006. Fast time scale dynamics of protein backbones: NMR relaxation methods, applications, and functional consequences. *Chem. Rev.* **106:** 1624–1671.

29. Igumenova, T., K. Frederick & A. Wand. 2006. Characterization of the fast dynamics of protein amino acid side chains using NMR relaxation in solution. *Chem. Rev.* **106:** 1672–1699.

30. Forman-Kay, J. 1999. The "dynamics" in the thermodynamics of binding. *Nat. Struct. Biol.* **6:** 1086–1087.

31. Cavanagh, J. & M. Akke. 2000. May the driving force be with you—whatever it is. *Nat. Struct. Biol.* **7:** 11–13.

32. Bosshard, H. 2001. Molecular recognition by induced fit: how fit is the concept. *News Physiol. Sci.* **16:** 171–173.

33. Kalodimos, C.G. 2011. NMR reveals novel mechanisms of protein activity regulation. *Protein Sci.* **20:** 773–782.

34. Marlow, M.S. J. Dogan, K.K. Frederick, *et al.* 2010. The role of conformational entropy in molecular recognition by calmodulin. *Nat. Chem. Biol.* **6:** 352–358.

35. Frederick, K.K., M.S. Marlow, K.G. Valentine, & A.J Wand. 2007. Conformational entropy in molecular recognition by proteins. *Nature* **448:** 325–329.

36. Hilser, V.J. 2010. Biochemistry. An ensemble view of allostery. *Science* **327:** 653–654.

37. Boehr, D.D., R. Nussinov & P.E. Wright. 2009. The role of dynamic conformational ensembles in biomolecular recognition. *Nat. Chem. Biol.* **5:** 789–796.

38. Goodey, N. & S. Benkovic. 2008. Allosteric regulation and catalysis emerge via a common route. *Nat. Chem. Biol.* **4:** 474–482.

39. Ma, B. & R. Nussinov. 2009. Amplification of signaling via cellular allosteric relay and protein disorder. *Proc. Natl. Acad. Sci. U.S.A.* **106:** 6887–6888.

40. Masterson, L.R. T. Yu, L. Shi, *et al.* 2011. cAMP-dependent protein kinase a selects the excited state of the membrane substrate phospholamban. *J. Mol. Biol.* **412:** 155–164.

41. Masterson, L., A. Mascioni, N. Traaseth, *et al.* 2008. Allosteric cooperativity in protein kinase A. *Proc. Natl. Acad. Sci. U.S.A.* **105:** 506–511.

42. Vendruscolo, M. 2011. Protein regulation: the statistical theory of allostery. *Nat. Chem. Biol.* **7:** 411–412.

43. Killian, B.J., J.Y. Kravitz, S. Somani, *et al.* 2009. Configurational entropy in protein-peptide binding: computational study of Tsg101 ubiquitin E2 variant domain with an HIV-derived PTAP nonapeptide. *J. Mol. Biol.* **389:** 315–335.

44. Homans, S. 2005. Probing the binding entropy of ligand-protein interactions by NMR. *Chembiochem* **6:** 1585–1591.

45. Sarkar, P., C. Reichman, T. Saleh, *et al.* 2007. Proline cis-trans isomerization controls autoinhibition of a signaling protein. *Mol. Cell.* **25:** 413–426.

46. Sarkar, P., T. Saleh, S-R. Tzeng, *et al.* 2011. Structural basis for regulation of the Crk signaling protein by a proline switch. *Nat. Chem. Biol.* **7:** 51–57.

47. Keramisanou, D., N. Biris, I. Gelis, *et al.* 2006. Disorder-order folding transitions underlie catalysis in the helicase motor of SecA. *Nat. Struct. Mol. Biol.* **13:** 594–602.

48. Gelis, I., A.M.J.J. Bonvin, D. Keramisanou, *et al.* 2007. Structural basis for signal-sequence recognition by the translocase motor SecA as determined by NMR. *Cell* **131:** 756–769.

49. Religa, T.L., R. Sprangers & L.E Kay. 2010. Dynamic regulation of archaeal proteasome gate opening as studied by TROSY NMR. *Science* **328:** 98–102.

50. Henzler-Wildman, K. & D. Kern. 2007. Dynamic personalities of proteins. *Nature* **450:** 964–972.

Ann. N.Y. Acad. Sci. ISSN 0077-8923

ANNALS OF THE NEW YORK ACADEMY OF SCIENCES

Issue: *Blavatnik Awards for Young Scientists*

Plate tectonics and planetary habitability: current status and future challenges

Jun Korenaga

Department of Geology and Geophysics, Yale University, New Haven, Connecticut

Address for correspondence: Jun Korenaga, Department of Geology and Geophysics, Yale University, P.O. Box 208109, New Haven, CT 06520-8109. jun.korenaga@yale.edu

Plate tectonics is one of the major factors affecting the potential habitability of a terrestrial planet. The physics of plate tectonics is, however, still far from being complete, leading to considerable uncertainty when discussing planetary habitability. Here, I summarize recent developments on the evolution of plate tectonics on Earth, which suggest a radically new view on Earth dynamics: convection in the mantle has been speeding up despite its secular cooling, and the operation of plate tectonics has been facilitated throughout Earth's history by the gradual subduction of water into an initially dry mantle. The role of plate tectonics in planetary habitability through its influence on atmospheric evolution is still difficult to quantify, and, to this end, it will be vital to better understand a coupled core–mantle–atmosphere system in the context of solar system evolution.

Keywords: terrestrial planets; mantle dynamics; planetary magnetism; atmospheric evolution

Introduction

Under what conditions can a planet like Earth—that is, a planet that can host life—be formed? This question of planetary habitability has been addressed countless times in the past,[1–3] as it is deeply connected to the origin of life, perhaps the most fascinating problem in science. In the last decade or so, research activities in this field have been invigorated, fueled by a rapidly expanding catalog of extrasolar planets.[4–6] The habitability of a planet depends on a number of factors including, for example, the mass of the central star and the distance from it, the atmospheric composition, orbital stability, the operation of plate tectonics, and the acquisition of water during planetary formation. The mass of the star determines the evolution of its luminosity, and the heliocentric distance of a planet as well as the volume of the atmosphere and its composition then control the surface temperature of the planet. The surface temperature has to be in a certain range so that we can expect the presence of liquid water provided that water exists, and orbital dynamics affect the stability of the planetary climate. Plate tectonics controls the evolution of the atmosphere

through volcanic degassing and subduction, and it is also essential for the existence of a planetary magnetic field, which protects the atmosphere from the interaction with the solar wind. These factors affecting planetary habitability are thus interrelated to various degrees. Whether or not plate tectonics is operating on a planet, for example, would give rise to vastly different scenarios for its atmospheric evolution, affecting the definition of the habitable heliocentric distance, that is, the habitable zone.

The focus of this contribution is on plate tectonics. Plate tectonics refers to a particular mode of convection in a planetary mantle, which is made of silicate rocks, and so far it is observed only on Earth. Earth's surface is divided into a dozen plates or so, and these plates are moving at different velocities. Most geological activities, such as earthquakes, volcanic eruption, and mountain building, occur when different plates interact at plate boundaries. The realization that Earth's surface is actively deforming via plate tectonics was achieved through the 1960s and 1970s, revolutionizing almost all branches of earth sciences. Plate tectonics is a fundamental process, yet we still do not understand it in a satisfactory manner. For example, whereas the present-day

doi: 10.1111/j.1749-6632.2011.06276.x

plate motion is known in considerable detail,[7] reconstructing past plate motion becomes quite difficult once we enter the Precambrian (before 540 million years ago), and even the gross characteristics of ancient plate tectonics is uncertain.[8] Naturally, when plate tectonics started to operate on Earth is still controversial.[9] Part of the difficulty originates in the paucity of observations; we have fewer geological samples from greater ages. The situation is even more compounded by the lack of theoretical understanding. Geophysicists have yet to form a consensus on why plate tectonics takes place on Earth and not on other terrestrial (e.g., Earth-like) planets such as Venus and Mars.[10] The physics of plate tectonics is still incomplete, and this creates a serious impediment to the discussion of planetary habitability. Under what conditions could plate tectonics emerge on a planet, and how would it evolve through time? Without being able to answer these questions, it would be nearly impossible to predict the atmospheric evolution of a given planet and thus its habitability.

Fundamental issues regarding the physics of plate tectonics may be paraphrased by the following questions: how did plate tectonics evolve in the past?, why does plate tectonics take place on Earth?, and when did plate tectonics first appear on Earth? Considerable progress has been made on the first question in the last decade, and this progress turns out to help better address the second and third questions as well. In the following sections, I will review each question one by one and conclude with a synthesis of current status as well as major theoretical challenges to be tackled in the coming years.

How did plate tectonics evolve?

As plate tectonics is just one type of thermal convection, it is reasonable to speculate on the evolution of plate tectonics on the basis of fluid mechanics. Earth's mantle in the past was generally hotter and thus probably had lower viscosity than present. Elementary fluid mechanics tells us that this reduction in viscosity should have resulted in more vigorous convection, that is, higher heat flux and faster plate tectonics.[11] Geological support for such faster plate tectonics has long been lacking,[12] but this lack of observational support is usually not taken seriously because geological data become very scarce and more difficult to interpret in the Precambrian. The notion of faster plate tectonics in the past,

however, has been known to predict an unrealistic thermal history called "thermal catastrophe,"[13] unless one assumes that Earth contains considerably more heat-producing elements than the composition models of Earth indicate (Fig. 1). This can be understood by considering the following global heat balance:

$$C\frac{dT}{dt} = H(t) - Q(t), \qquad (1)$$

where C is the heat capacity of the entire Earth, T is average internal temperature, t is time, H is internal heat production owing to the decay of radioactive isotopes, and Q is heat loss from the surface by mantle convection. By the nature of radioactive decay, the internal heat production monotonically

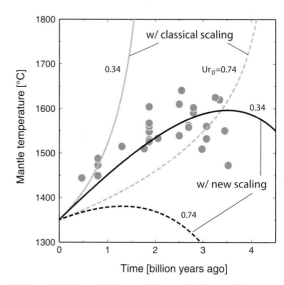

Figure 1. Thermal history prediction for four combinations of heat flow scaling and internal heat production (see Ref. 17 for modeling detail). The new scaling of plate tectonics predicts relatively constant heat flux independent of mantle temperature, whereas classical scaling predicts higher heat flux for hotter mantle. The Urey ratio is a measure of the amount of heat-producing elements in the mantle, and the chemical composition models of Earth suggest that its present-day value ($Ur_0 = H(0)/Q(0)$) is relatively low, ~0.3.[14] Constant heat flux with a low present-day Urey ratio (*solid*) is the only one that can reproduce the observed concave-downward thermal history with an average cooling rate of ~100 K Ga⁻¹ (*circles*).[27] In this prediction, Earth was warming up during the first one billion years; such a situation is possible with the efficient cooling of the magma ocean.[50] Classical scaling with a low Urey ratio results in thermal catastrophe (*gray line*). Classical scaling with a high Urey ratio (*gray dashed line*) can reproduce a reasonable cooling rate, but a thermal history is concave upward. Constant heat flux with a high Urey ratio (*dashed line*) results in too cold a thermal history.

decreases with time, with an effective half-life of about three billion years. The convective heat loss can be parameterized as a function of average temperature as

$$Q \propto T^m, \qquad (2)$$

where the exponent m is predicted to be $\sim$10 by the classical theory of thermal convection;[11] heat loss is extremely sensitive to a change in internal temperature. The present-day internal heat production $H(0)$ is only about 30% of the convective heat loss $Q(0)$,[14] so about 70% of heat loss must be balanced by the rapid cooling of Earth; that is, Earth must have been much hotter than at present to explain the present-day thermal budget. Because convective heat loss rises sharply with increasing temperature (Eq. 2), however, heat loss must have been extremely high in the past, resulting in a more severe imbalance between heat production and heat loss, as we consider further back in time. This positive feedback is what causes thermal catastrophe in the middle of the Earth history. The only way to prevent it, while keeping the classical scaling ($m \sim 10$), is to assume that internal heat production is close to convective heat loss at present, that is, $H(0) \sim Q(0)$, but this violates our understanding of the chemical budget of Earth. This conflict between the geophysical theory of mantle convection and the geochemical model of Earth has inspired a variety of proposals (see Ref. 14 for review), many of which hide an excessive amount of heat-producing elements in the deep, inaccessible mantle—a possible but rather *ad hoc* solution.

A novel solution was suggested in 2003 based on the effect of mantle melting on mantle convection.[15] Faster plate tectonics in the past is based on simple fluid mechanics that do not capture realistic complications associated with silicate rocks. An important difference from classical thermal convection is chemical differentiation; when the mantle is rising toward the surface, it usually melts, and this melting can affect mantle dynamics. Upon melting, impurities in the mantle, most notably water, are largely partitioned into the melt phase, leaving the residual mantle very stiff.[16] A hotter mantle in the past means more extensive melting, making thicker stiff plates and slowing down plate tectonics. Considering both the physics and chemistry of Earth's mantle thus points to an entirely opposite prediction: slower plate tectonics in the past, which

is equivalent to using $m \leq 0$ in Eq. (2). With this nonclassical scaling of plate tectonics, it has become possible to reconstruct a reasonable thermal history without violating the geochemical constraints (Fig. 1). This solution, which was further elaborated in 2006,[17] met considerable skepticism because of its counterintuitive nature. Some doubted the robustness of the geochemical constraints on the amount of heat-producing elements, but the uncertainty of the mantle composition has been shown to be tight enough to discount such leeway.[18] Others suspected that the relative contribution of heat-producing elements may be increased by lowering the estimate on present-day heat flux instead,[19,20] but this possibility has been shown to be inconsistent with available geological records.[21,22] Additionally, the counterintuitive prediction was based on an approximate theory (known as the boundary layer theory) with several simplifying assumptions, and some questioned the validity of this approach.[23] Recently, however, the original prediction has been given full theoretical support from extensive numerical simulation and scaling analysis.[24,25]

Equally important is the appearance of new decisive observations. In 2008, the compilation of the geological records of ancient passive margins was published, which indicates that the tempo of plate tectonics in the past was indeed slower than present.[26] In 2010, the thermal history of Earth's upper mantle was reconstructed by applying the latest petrological technique to an extensive compilation of Precambrian volcanic rocks (Fig. 1).[27] The concave-downward nature of this thermal history is particularly important, as it provides strong support for the notion of slower plate tectonics and the relatively low abundance of heat-producing elements at the same time; it is impossible to reproduce this curvature by assuming faster plate tectonics for a hotter mantle. Most recently, a new constraint on the abundance of heat-producing elements in the mantle was reported based on geoneutrino observations, which favors the relatively low abundance as indicated by the geochemical estimate.[28]

The radically new view on the evolution of plate tectonics, therefore, has been corroborated both theoretically and observationally in recent years, and it has become difficult to refute the notion of slower plate tectonics in the past, however counterintuitive it might be. Actually, what is counterintuitive is a subjective matter, and in this case, it is largely educational. Faster plate tectonics for a hotter Earth is predicted by the fluid mechanics of a nearly iso-viscous fluid. No theoretical justification exists for its applicability to Earth's mantle. The classical theory is still widely used in planetary sciences, but it simply fails to reproduce the thermal history of the best-understood planet (Fig. 1). There would be little merit in extrapolating a theory that cannot explain Earth to other terrestrial planets, for which we have considerably fewer observational constraints. This is especially true when discussing the dynamics of Earth-like, potentially habitable planets.

Why does plate tectonics happen?

There are two fundamentally different modes of mantle convection: (1) plate tectonics and (2) stagnant lid.[29] In stagnant-lid convection, the entire surface of a planet forms a rigid spherical shell, and convection can take place only under the shell. In plate tectonics, the surface is broken into pieces, most of which can return to the deep mantle, enabling geochemical cycles between the surface and the interior. Among the four terrestrial planets in our solar system, Earth is the only planet that exhibits plate tectonics, and the other three (Mercury, Venus, and Mars) are believed to be in the mode of stagnant lid.[30] It is easy to explain why plate tectonics does not take place on other planets, because stagnant-lid convection is the most natural mode of convection in a medium with strongly temperature-dependent viscosity, such as silicate rocks that constitute a planetary mantle.[29] Mantle viscosity is extremely high at a typical surface condition, so virtually no deformation is expected there. The mode of plate tectonics is possible, therefore, only when some additional mechanism exists to compensate the effect of temperature-dependent viscosity. Ongoing debates are mostly regarding this additional weakening mechanism.

In addition to ductile deformation characterized by viscosity, silicate rocks can also deform by brittle deformation such as cracking and faulting. Weakening by brittle mechanisms is limited by frictional strength,[31] however, and with a typical frictional coefficient of order 1, brittle weakening is insufficient to cause plate tectonics.[32] In order to simulate plate tectonics in numerical models, therefore, it has been a common practice to assume a much lower friction coefficient, but a physical mechanism that

could lead to such a low coefficient has been poorly understood.[33] One plausible mechanism is a reduction in an effective friction coefficient by high pore fluid pressure, with water being the fluid medium. For plate tectonics to occur with this mechanism (i.e., to break a thick stagnant lid), however, water has to be transported to substantial depths and then isolated from the surface to achieve high pore fluid pressure. The mere existence of surface water does not guarantee either of these requirements. If deep water is connected to the surface, for example, it would be at hydrostatic pressure, meaning that pore fluid pressure is too low to achieve a sufficiently low friction coefficient. In this regard, the thermal cracking hypothesis,[33] in which a rigid lid is extensively fractured by strong thermal stress and then later sealed by hydration reactions, has so far been the only tangible mechanism that could generate plate tectonics in the presence of surface water (Fig. 2).

When discussing the mode of mantle convection, it is important to avoid being trapped in a chicken-and-egg situation. The bending of a subducting plate, for example, may fracture and weaken the plate by hydration,[34] but one cannot invoke this mechanism for the onset of plate tectonics; a weakening mechanism must be operational even without plate tectonics. The same caution applies to various dynamic weakening mechanisms associated with earthquake dynamics.[35]

Finding a physical mechanism for weakening is just one side of the coin. The other side is to understand the critical strength of a surface lid that can be overcome by convective stress exerted by the mantle below. Both sides are necessary to understand under what conditions plate tectonics can happen. This issue has been studied by various authors using numerical simulation (e.g., Refs. 36 and 37), but in most previous attempts, the temperature dependency of mantle viscosity was not strong enough to be Earth-like. A quantitative criterion for the onset of plate tectonics was found in 2010, for the first time with realistic mantle viscosity, while conducting a number of numerical simulations to establish the scaling of plate tectonics for thermal evolution modeling.[24] Revisiting the notion of slower ancient plate tectonics with this new criterion turns out to yield an intriguing insight for the initiation of plate tectonics on Earth, as discussed in the next section.

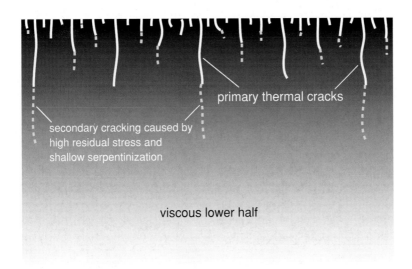

Figure 2. Schematic illustration for rheological evolution within a plate under oceans.[33] Optimal release of thermal stress developed in a cooling plate is achieved by a cascade crack system (primary cracks). Any residual stress will eventually be released by secondary crack propagation if partial crack healing by shallow serpentinization raises the pressure of trapped seawater to lithostatic pressure. The stiffest part of plate due to strong temperature-dependent viscosity can thus be pervasively weakened by thermal cracking and subsequent hydration.

When did plate tectonics start?

Based on field evidence, many geologists would concur with the operation of plate tectonics back to about 3 billion years ago,[38,39] but anything beyond that is controversial. Earth's history is divided into four eons: the Hadean (4.6–4.0 billion years ago), the Archean (4.0–2.5 billion years ago), the Proterozoic (2.5–0.54 billion years ago), and the Phanerozoic (0.54 billion years ago to present). The Hadean–Archean boundary is defined by the age of the oldest rock found on Earth. The Archean–Proterozoic boundary is defined by the relative abundance of rocks—that is, rocks of Archean ages are much rarer than those of younger ages. These definitions of the geological time scale indicate that finding unambiguous geological data for the first appearance of plate tectonics on Earth's history, which might be in the Hadean era,[40] would be quite challenging. Building a theoretical foundation for this problem is thus of critical importance.

As mentioned earlier, slower plate tectonics in the past results from the formation of thicker stiff plates by more extensive melting. Thicker plates slow down plate tectonics, but if too thick, they could potentially jeopardize the operation of plate tectonics itself. The likelihood of shutting down plate tectonics in the past can be high because much lower convective stress is expected for a hotter, less viscous mantle beneath plates; thicker plates and weaker convective stress both act to impede plate tectonics. Indeed, a quantitative assessment of this possibility using the new criterion indicates that plate tectonics is viable only for the last one billion years,[41] which grossly contradicts with geological evidence.

One possible resolution to this conundrum came from an apparently unrelated thread of thought, though in hindsight it is a natural extension of the existing theory. By incorporating geological constraints on the past sea level into thermal evolution modeling, one can reconstruct the history of ocean volume, and slower plate tectonics corresponds to greater ocean volume in the past.[42] The Archean oceans are estimated to have been about twice as voluminous as the present oceans, and this difference in ocean volume is roughly equivalent to the amount of water stored in the present mantle. Earth's mantle could have been drier and thus more viscous in the past, and if this exchange of water between the mantle and the oceans is taken into account, the operation of plate tectonics becomes viable throughout Earth's history.[24,41] It has long been suggested that plate tectonics could result in net water influx to the mantle,[43,44] and modeling the mantle as an open system now appears to be a necessity, rather than an option.

Plate tectonics, therefore, could have started on Earth shortly after the solidification of a global magma ocean, which probably existed only for the first few tens of millions of years of Earth's history.[45] One interesting finding from Earth's evolution with a hydrating mantle is that the subduction of water is essential to maintain a dry land mass, which in turn plays an indispensable role in stabilizing the climate. Without a dry land mass, no silicate weathering could take place so that the atmospheric composition could not be regulated efficiently by carbon cycle.[1]

Summary and outlook

An emerging view on the evolution of plate tectonics on Earth can be summarized by the following. Plate tectonics started probably in the very early Earth, shortly after the solidification of the putative magma ocean. The onset of plate tectonics was facilitated by an initially dry mantle, which has since been slowly hydrated by plate tectonics. While Earth has been cooling down, plate tectonics has been speeding up, instead of slowing down. This is because a colder mantle leads to thinner, easily deformable plates and because the effect of hydration on viscosity tends to cancel the effect of temperature. This scenario has been shown to be internally consistent and dynamically plausible by the scaling laws of plate tectonics. Though being unconventional in nearly all aspects, it is the only hypothesis that is consistent with all of major observations relevant to Earth's evolution, including petrological constraints on the thermal history, geochemical constraints on the thermal budget, and geological constraints on the tempo of plate tectonics, the mode of mantle convection, and the global sea level change.

Water is thus expected to play fundamental roles in the initiation of plate tectonics and its evolution over Earth's history. The physics of elementary processes involving water in the above scenario, however, still requires considerable future development. The plausibility of the thermal cracking hypothesis, for example, needs to be tested further by modeling the physicochemical evolution of a multiple crack system. The hypothesis has indirect observational support through its impact on effective thermal expansivity,[46,47] but more direct evidence may be obtained by large-scale field experiments. Additionally, the rate of water transport to the deep mantle by subduction should be quantified from first principles. Though it is a highly complex problem involving petrology, mineral physics, and fluid mechanics, its solution is essential for a theory with predicting power.

If a terrestrial planet starts out with a dry mantle and surface water, which appears to be a likely initial condition for subsolidus mantle convection,[3] the onset of plate tectonics is probably justifiable. Predicting its subsequent evolution is, however, still a formidable task. One of the major uncertainties is the amount of surface water. On the basis of the planetary formation theory, the origin of Earth's water is often considered to be in the outer solar system,[48] and the delivery of water is a highly stochastic process. The quantity of water to be delivered does not have to be large; one ocean worth of water corresponds to only 0.02% of Earth's mass. A difficult part is how to maintain it over the geological time. The existence of a planetary magnetic field, which could provide a shield against solar wind erosion,[49] depends on the rate of core cooling. Earth's thermal history indicates that the mantle was warming up in the early Earth (Fig. 1). The core could still have been cooling during that time if the core was initially superheated by its formation process, but to answer whether the cooling rate was sufficient to drive a planetary dynamo, modeling the thermal evolution of a coupled core–mantle system would be critical. Understanding the atmospheric evolution of a given terrestrial planet, or its habitability at large, therefore, requires us to create a unified theoretical framework that spans from the solar system evolution to the dynamics of planetary interior.

Acknowledgments

This work was sponsored by a Microsoft A. Richard Newton Breakthrough Research Award. The author thanks Norm Sleep for a constructive review.

Conflicts of interest

The author declares no conflicts of interest.

References

1. Kasting, J.F. & D. Catling. 2003. Evolution of a habitable planet. *Annu. Rev. Astron. Astrophys.* **41:** 429–463.

2. Gaidos, E., B. Deschenes, L. Dundon, *et al.* 2005. Beyond the principle of plentitude: a review of terrestrial planet habitability. *Astron. J.* **5**: 100–126.

3. Zahnle, K., N. Arndt, C. Cockell, *et al.* 2007. Emergence of a habitable planet. *Space Sci. Rev.* **129**: 35–78.

4. Marcy, G.W. & R.P. Butler. 1998. Detection of extrasolar giant planets. *Annu. Rev. Astron. Astrophys.* **36**: 57–97.

5. Rivera, E.J., J.J. Lissauer, R.P. Butler, *et al.* 2005. A ~7.5 M-circle plus planet orbiting the nearby star, GJ 876. *ApJ* **634**: 625–640.

6. Borucki, W.J. *et al.* 2011. Characteristics of planetary candidates observed by Kepler, II: analysis of the first four months of data. *ApJ.* **736**: 19, doi: 10.1088/0004-637X/736/1/19.

7. DeMets, C., R.G. Gordon & D.F. Argus. 2010. Geologically current plate motions. *Geophys. J. Int.* **181**: 1–80.

8. Bleeker, W. 2003. The late Archean record: a puzzle in ca. 35 pieces. *Lithos* **71**: 99–134.

9. Condie, K.C. & V. Pease, Eds. 2008. *When Did Plate Tectonics Begin on Planet Earth?* Geological Society of America.

10. Bercovici, D. 2003. The generation of plate tectonics from mantle convection. *Earth Planet Sci. Lett.* **205**: 107–121.

11. Schubert, G., D. Stevenson & P. Cassen. 1980. Whole planet cooling and the radiogenic heat source contents of the earth and moon. *J. Geophys. Res.* **85**: 2531–2538.

12. Kröner, A. & P.W. Layer. 1992. Crust formation and plate motion in the early Archean. *Science* **256**: 1405–1411.

13. Christensen, U.R. 1985. Thermal evolution models for the Earth. *J. Geophys. Res.* **90**: 2995–3007.

14. Korenaga, J. 2008. Urey ratio and the structure and evolution of Earth's mantle. *Rev. Geophys.* 46:RG2007, doi:10.1029/2007RG000241.

15. Korenaga, J. 2003. Energetics of mantle convection and the fate of fossil heat. *Geophys. Res. Lett.* **30**: 1437, doi:10.1029/2003GL016982.

16. Hirth, G. & D.L. Kohlstedt. 1996. Water in the oceanic mantle: implications for rheology, melt extraction, and the evolution of the lithosphere. *Earth and Planetary Science Letters* **144**: 93–108.

17. Korenaga, J. 2006. Archean geodynamics and the thermal evolution of Earth. In *Archean Geodynamics and Environments*. K. Benn, J.-C. Mareschal & K. Condie, Eds.: 7–32. American Geophysical Union, Washington, D.C.

18. Lyubetskaya, T. & J. Korenaga. 2007. Chemical composition of Earth's primitive mantle and its variance, 1, methods and results. *J. Geophys. Res.*, **112**: B03211, doi:10.1029/2005JB004223.

19. Grigne, C., S. Labrosse & P.J. Tackley. 2005. Convective heat transfer as a function of wavelength: implications for the cooling of the Earth. *J. Geophys. Res.* **110**: B03409, doi:10.1029/2004JB003376.

20. Stevenson, D.J. 2007. A planetary scientist foresees a shift in the debate about Earth's heat flow. *Nature* **448**: 843.

21. J. Korenaga. 2007. Eustasy, supercontinental insulation, and the temporal variability of terrestrial heat flux. *Earth Planet Sci. Lett.* **257**: 350–358.

22. Loyd, S.J., T.W. Becker, C.P. Conrad, *et al.* 2007. Time variability in Cenozoic reconstructions of mantle heat flow: plate tectonic cycles and implications for Earth's thermal evolution. *Proc. Nat. Acad. Sci. USA* **104**: 14266–14271.

23. Davies, G.F. 2009. Effect of plate bening on the Urey ratio and the thermal evolution of the mantle. *Earth Planet Sci. Lett.* **287**: 513–518.

24. Korenaga. J. 2010. Scaling of plate-tectonic convection with pseudoplastic rheology. *J. Geophys. Res.*, **115**: B11405, doi:10.1029/2010JB007670, 2010.

25. Rose I.R. & J. Korenaga. 2011. Mantle rheology and the scaling of bending dissipation in plate tectonics. *J. Geophys. Res.* **116**: B06404, doi:10.1029/2010JB008004.

26. Bradley, D.C. 2008. Passive margins through earth history. *Earth Sci. Rev.* **91**: 1–26.

27. Herzberg, C., K. Condie & J. Korenaga. 2010. Thermal evolution of the Earth and its petrological expression. *Earth Planet Sci. Lett.* **292**: 79–88.

28. Gando, A. *et al.* 2011. Partial radiogenic heat model for Earth revealed by geoneutrino measurements. *Nature Geosci.* **4**: 647–651, doi:10.1038/ngeo1205.

29. Solomatov, V.S. 1995. Scaling of temperature- and stress-dependent viscosity convection. *Phys. Fluids* **7**: 266–274.

30. Schubert, G., D.L. Turcotte & P. Olson. 2001. *Mantle Convection in the Earth and Planets*. Cambridge University Press, Cambridge, United Kingdom.

31. Scholz, C.H. 2002. *The Mechanics of Earthquakes and Faulting*. Cambridge University Press, Cambridge, United Kingdom.

32. Moresi, L. & V. Solomatov. 1998. Mantle convection with a brittle lithosphere: thoughts on the global tectonic styles of the Earth and Venus. *Geophys. J. Int.* **133**: 669–682.

33. Korenaga. J. 2007. Thermal cracking and the deep hydration of oceanic lithosphere: a key to the generation of plate tectonics? *J. Geophys. Res.*, **112**: B05408, doi:10.1029/2006JB004502.

34. Ranero, C.R., J. Phipps Morgan, K. McIntosh & C. Reichert. 2003. Bending-related faulting and mantle serpentinization at the middle America trench. *Nature* **425**: 367–373.

35. Korenaga, J. 2010. On the likelihood of plate tectonics on super-Earths: does size matter? *ApJ* **725**: L43–L46.

36. Richards, M.A., W.-S. Yang, J.R. Baumgardner & H.-P. Bunge. 2001. Role of a low-viscosity zone in stabilizing plate tectonics: implications for comparative terrestrial planetology. *Geochem. Geophys. Geosys.* **2**: 2000GC000115.

37. Stein, C., J. Schmalzl & U. Hansen. 2004. The effect of rheological parameters on plate behavior in a self-consistent model of mantle convection. *Phys. Earth Planet Inter.* **142**: 225–255.

38. Hoffman, P.F. 1997. Tectonic genealogy of North America. In *Earth Structure: An Introduction to Structural Geology and Tectonics*. B.A. van der Pluijm & S. Marshak, Eds.: 459–464. McGraw-Hill, New York.

39. Van Kranendonk, M.J., R.H. Smithies, A.H. Hickman & D.C. Champion. 2007. Review: secular tectonic evolution of Archean continental crust: interplay between horizontal and vertical processes in the formation of the Pilbara Craton, Australia. *Terra Nova* **19**: 1–38.

40. Hopkins, M.D., T.M. Harrison & C.E. Manning. 2010. Constraints on Hadean geodynamics from mineral inclusions in >4 Ga zircons. *Earth Planet Sci. Lett.* **298**: 367–376.

41. Korenaga, J. Thermal evolution with a hydrating mantle and the initiation of plate tectonics in the early Earth. *J. Geophys. Res.* doi:10.1029/2011JB008410, in press.

42. Korenaga, J. 2008. Plate tectonics, flood basalts, and the evolution of Earth's oceans. *Terra Nova* **20:** 419–439.

43. Ito, E., D.M. Harris & A.T. Anderson. 1983. Alteration of oceanic crust and geologic cycling of chlorine and water. *Geochim. Cosmochim. Acta* **47:** 1613–1624.

44. Smyth, J.R. & S.D. Jacobsen. 2006. Nominally anhydrous minerals and Earth's deep water cycle. In *Earth's Deep Water Cycle*. S.D. Jacobsen & S. van der Lee, Eds.: 1–11. American Geophysical Union.

45. Solomatov, V. 2007. Magma oceans and primordial mantle differentiation. In *Treatise on Geophysics*. Vol. **9**, G. Schubert, Ed.: 91–119. Elsevier, Amsterdam.

46. Korenaga, J. 2007. Effective thermal expansivity of Maxwellian oceanic lithosphere. *Earth Planet Sci. Lett.* **257:** 343–349.

47. Korenaga, T. & J. Korenaga. 2008. Subsidence of normal oceanic lithosphere, apparent thermal expansivity, and seafloor flattening. *Earth Planet Sci. Lett.* **268:** 41–51.

48. Morbidelli, A., J. Chambers, J.I. Lunine, *et al.* 2000. Source regions and timescales for the delivery of water to the Earth. *Meteorit. Planet Sci.* **35:** 1309–1320.

49. Lundin, R., H. Lammer & I. Ribas. 2007. Planetary magnetic fields and solar forcing: implications for atmospheric evolution. *Space Sci. Rev.* **129:** 245–278.

50. Sleep, N.H. 2000. Evolution of the mode of convection within terrestrial planets. *J. Geophys. Res.* **105:** 17563–17578.

Ann. N.Y. Acad. Sci. ISSN 0077-8923

ANNALS OF THE NEW YORK ACADEMY OF SCIENCES

Issue: *Blavatnik Awards for Young Scientists*

Accurate evaluation and analysis of functional genomics data and methods

Casey S. Greene[1] and Olga G. Troyanskaya[1,2]

[1]Lewis-Sigler Institute for Integrative Genomics, Princeton University, Princeton, New Jersey. [2]Department of Computer Science, Princeton University, Princeton, New Jersey

Address for correspondence: Olga G. Troyanskaya, Lewis-Sigler Institute for Integrative Genomics, 242 Carl Icahn Laboratory, Princeton University, Princeton, NJ 08544. ogt@genomics.princeton.edu

The development of technology capable of inexpensively performing large-scale measurements of biological systems has generated a wealth of data. Integrative analysis of these data holds the promise of uncovering gene function, regulation, and, in the longer run, understanding complex disease. However, their analysis has proved very challenging, as it is difficult to quickly and effectively assess the relevance and accuracy of these data for individual biological questions. Here, we identify biases that present challenges for the assessment of functional genomics data and methods. We then discuss evaluation methods that, taken together, begin to address these issues. We also argue that the funding of systematic data-driven experiments and of high-quality curation efforts will further improve evaluation metrics so that they more-accurately assess functional genomics data and methods. Such metrics will allow researchers in the field of functional genomics to continue to answer important biological questions in a data-driven manner.

Keywords: bioinformatics; functional genomics; evaluation

Introduction

With the past decade's explosion of genome-scale biomedical data, innovative computing approaches in genomics can revolutionize our understanding of biology and medicine and provide unprecedented levels of public access to biomedical information.[1] Genomics experiments now not only provide a "parts list" of the organism in the form of its genomic sequence, but also can assess what these genes do, how they are controlled in cellular pathways, and what malfunctions in these cellular systems lead to disease.[2]

However, despite this explosion of high-throughput data in molecular biology, our functional understanding of cellular processes remains incomplete. This gap between data generation and the discovery of reliable functional information is largely due to the lack of specificity and resolution in high-throughput data. In other words, the challenge lies in identifying the true biological signal and separating it from both technical and experimental noise. To alleviate this lack of specificity and extract accurate functional information from these data, integrated analysis of heterogeneous data sources with robust computational methods is necessary.[3] Ideally, such analyses can leverage multiple, often complementary, sources of information in diverse biological datasets to generate precise biological hypotheses about what proteins do (gene function),[4–6] how they do it (interactions and regulation),[7,8] and how various perturbations (i.e., disease or drugs) affect biological pathways.[9,10]

The noisy and heterogeneous nature of functional genomic data and the lack of accurate gold standards, or biological "truths" that can be used to assess accuracy and coverage of each dataset, make it challenging to analyze these data and generate specific, experimentally testable hypotheses. Even the best algorithms can be stymied by significant evaluation biases that can lead to trivial or incorrect predictions with apparently higher accuracy. These biases are often highly technical and data dependent, yet their understanding is critical to any

doi: 10.1111/j.1749-6632.2011.06383.x

Ann. N.Y. Acad. Sci. 1260 (2012) 95–100 © 2012 New York Academy of Sciences.

analysis of these diverse functional genomics data, be it for prediction of protein function and interactions, or for more complex modeling tasks, such as building biological pathways. Below, we discuss the major sources of these biases, and propose ways to detect and avoid them when dealing with functional genomics datasets. When these biases are properly addressed, functional genomics approaches have great potential to make high-quality predictions of gene function, disease involvement, or tissue specificity. These predictions can be used to efficiently and effectively target genes for further experimental study.

Biases in analysis and evaluation of functional genomics data

The biases that affect the analysis and evaluation of functional genomics data can arise from biological factors and from the actual process of measuring biological systems. Here, we discuss four such biases, which we call process bias, term bias, standard bias, and annotation distribution bias.

Process bias can arise when distinct biological groups of genes, processes, or functions, such as the cell cycle DNA damage repair, are grouped for evaluation. A classic example of this comes from the ribosome pathway in yeast. Ribosomes are the organelles in the cell responsible for translating mRNA to protein, and their components are very easily measured by a specific experimental platform called gene expression microarrays, which comprise the majority of existing functional genomics data. Myers et al.[11] showed that, when evaluation is performed on the grouped set of processes, a single easy-to-predict process, such as the ribosome, can dramatically alter the evaluation results (Fig. 1). In the case of the ribosome, this difference is so pronounced that the gene coexpression data, which is clearly the best overall predictor when the ribosome is grouped with other processes, is only of average utility once the ribosome pathway is removed from a set of 99 pathways used for evaluation.

Term bias arises when the gold standard is correlated with other factors. At its most extreme, this can occur because of the direct inclusion of evaluation standards in data used for training, such as inclusion of one curated database (for example, Gene Ontology)[12,13] as input data when another (for example, KEGG)[14] is used for evaluation. In functional genomics data, these term biases can also oc-

cur through subtle contamination. One example of this hard-to-detect yet significant bias comes from the presence or absence of genes in large-scale genomic datasets. Because genes in most large-scale datasets are not randomly selected (for example, genes measured by microarray platforms), genes with more associated terms are more likely to be well represented. To a researcher who assumes that the presence or absence of genes in the data is random, using the sum of a score for a gene would not alter performance from using the mean score. In practice, however, there is a term bias related to which genes are well represented. Therefore, the sum appears dramatically better than the mean, but this is only because the sum acts as a proxy for how well studied the gene is. In this way, small algorithmic differences can lead to large performance changes when they exploit subtle, or even hidden, contamination between the training and evaluation sets.

Standard bias can impact evaluation methods because, in the traditional biological literature, biologists do not randomly select genes for the study. Researchers studying a specific phenotype will select genes that they expect to be involved in the process for assay. This selection step is highly nonrandom, and can create an unexpected and difficult to assess bias. Huttenhower et al.[15] found that in the case of mitochondrial biogenesis, while methods' ability to predict gene–phenotype relationships assessed through cross-validation differed, the actual ability of the methods to predict novel relationships (verified through experimental study) was very similar (Fig. 2). Furthermore, Hibbs et al.[16] observed that existing annotations of genes to phenotypes are often biased toward very severe gene–phenotype relationships and that subtle phenotypes are underrepresented. These two results, taken together, show how the process by which experiments are performed in molecular biology and genetics can introduce subtle biases into standards that mislead even very careful evaluation. These results also highlight the potential of functional genomics. By providing high-quality predictions from large-scale data compendia, a targeted experimental study based on carefully designed functional genomics approaches can reduce the bias in our existing knowledge.

The final challenge that we discuss is *annotation distribution bias*. This occurs because genes are not evenly annotated to functions and phenotypes. This broad size distribution creates difficulties in the

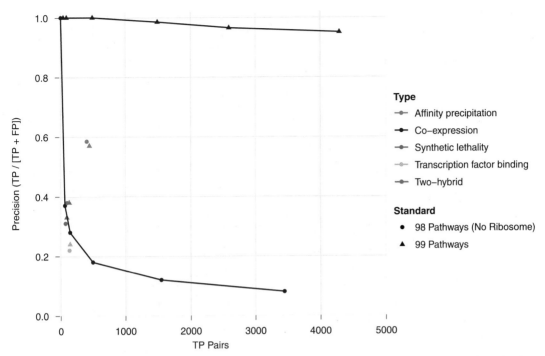

Figure 1. The performance of microarray coexpression data (the blue line) appears to be by far the best predictor (quality is indicated by precision at the specified number of true positive pairs) when the ribosome is included among the evaluation pathways (triangle symbols). When the ribosome is removed from consideration (circle symbols), coexpression is not substantially better than the other types of data.

assessment of functional genomics data and methods.[17] The best predictions for biological follow-up are both highly specific and accurate, but the best predictions from a performance standpoint (as performance is generally measured) are very broad functions or phenotypes. This is because predictions to these broad terms are more likely to be accurate by chance alone. We have set up a web server at http://dfp.princeton.edu that illustrates this effect. The web server predicts future annotations to the Gene Ontology (GO),[12,13] a common source of functional annotations, based on existing annotations as of an "evaluation cutoff" date. The predictions are made solely based on the number of existing annotations to each term, and thus use no gene-specific information (in other words, this is "data-free prediction"). The performance of this method is competitive with modern prediction methodologies.[5] This indicates that performance on standard metrics does not necessarily reflect the actual utility of functional genomics data or methods, necessitating careful correction for such biases.

Avoiding and addressing evaluation biases

Despite these challenges, it is possible to perform meaningful evaluation of functional genomics data and methods through careful and critical assessment. There are computational solutions that, when used with care, address these challenges and make accurate and unbiased evaluation possible. There are also ways to effectively use additional experimental data to supplement computation analyses. Finally, computationally directed, comprehensive experimental follow-up can be performed. Such systematically directed experiments are the ideal, though often too costly in time and money, solution, as they address all of the above concerns by providing direct experimental confirmation of results.

Individual computational assessments each have weaknesses that can compromise their utility, but used carefully and taken together they can provide a great deal of information about the effectiveness of functional genomics methods and data. To avoid process bias, distinct processes should be evaluated

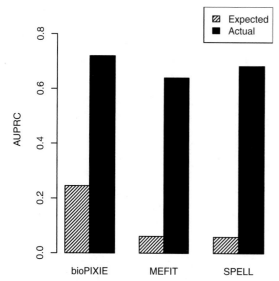

Figure 2. The expected AUPRC from cross-validation (diagonally hashed) and actual performance during experimental validation (solid) of various functional genomics strategies for gene-function prediction. The fold change between the expected and actual performance of the integrative method (bioPIXIE) is dramatically different than that for the coexpression methods MEFIT and SPELL. The expected performance can be influenced by subtle biases, so careful analysis is required to insure that results will generalize to actual performance.

data and methods should take into account the utility of predictions, either through the annotation of expert biologists[11] or metrics of prediction specificity.[24]

Standard bias is the most difficult to address. One way of avoiding standard bias is by making a final performance assessment through a blinded literature review. Because there are not sufficient professional curators to fully catalog all experimental results for gold standard datasets, often unannotated examples still remain in the literature. Data-driven methods can then discover the same phenotype, function, or tissue relationship. A careful analysis of the literature can then discover these underannotations, which can be used as a benchmark in a hand-curated or even in an automated manner. For each function of interest, genes from the function are paired with randomly selected genes. This set of genes is shuffled and provided in an unlabeled form. These genes are then evaluated for their role in the function of interest based on currently available literature. After genes are classified based by literature evidence, the frequency of literature support for predictions from the method or data to those chosen randomly provides an assessment of the quality of the data or method.

Of course, the most definitive way to assess functional genomics data and methods is by the biological validation of predictions. This can be done as part of a predefined experimental pipeline. For example, Hess *et al.* predicted and validated more than 100 proteins that affect the biogenesis and inheritance of mitochondria in *Saccharomyces cerevisiae*.[25] Interestingly, the computationally directed experimental approach allowed for the discovery of more genes with subtle mitochondrial phenotypes than had previously been observed.[16] Practical and cost concerns dictate that this can only occur in a small subset of studies, but when feasible, this provides the highest quality assessment. This evaluation can also be done on a large multilab scale when sufficient funding is available. The COMBREX project now provides grants to biologists that agree to test prespecified computational predictions.[26] To insure the utility of such work, the evaluation data from such experiments (including both confirmed and unconfirmed predictions) should be freely shared with the community to provide high-quality evaluation metrics for other studies.

separately. If a single summary statistic is required, distinct functions should only be combined after insuring there are no outliers that will dramatically change the interpretation of results. If there are outliers, results should be provided with and without outliers, as in Myers *et al.*[11]

Term bias can be mitigated with the addition of a temporal holdout. As long as the functional genomics data are fixed on a certain day, and phenotype or function assignments after that day are used for evaluation, this metric should avoid many of the hidden circularity issues that can affect simple random holdouts. Many functional genomics studies use either a temporal or a random holdout,[18–23] but using both can help mitigate the common biases that we identify. It is also important to realize that not all correct predictions are uniformly useful. Many easy-to-predict terms are highly generic (for example protein phosphorylation), yet good evaluation numbers can be obtained relatively easily from such not-so-useful predictions. Instead, the analysis of functional genomics

Discussion

The difficulty of evaluating functional genomics data presents distinct challenges to the fields of functional genomics, but there are challenges that can be addressed through careful experimental design. The community needs to be very consistent and careful during evaluation to insure that neither those developing nor those using the algorithms or data are misled by incomplete evaluations. Publicly available gold standard sets derived from systematic high-quality experiments, such as those now available in mitochondria,[25] are very helpful for evaluation, but more are required. The COMBREX strategy of prioritizing experiments that evaluate functional genomics predictions may also help in this regard.[26]

Current standard computational evaluation methods often rely upon curated GO[13] or KEGG[14] annotations. These functional annotations are extremely useful in this regard, but are incomplete due to the breadth of modern biomedical and biological literature. More funding allocated for such expert curation is a cost-effective way to convert information from biomedical literature into well-structured knowledge. Expanding funding for human curators will expand the coverage of GO and other such resources, thereby broadening the applicability of evaluation metrics that use these sources. A complementary strategy to increase the proportion of structured knowledge uses structured publication[27,28] and next-generation bioinformatics tools like PILGRM[29] that allow biologists to direct the analysis of large data compendia. Because biologists use their own literature review to direct the analysis, they effectively curate a source of structured knowledge suitable for evaluation. In combination, these approaches will support more accurate and useful predictions, thus enabling faster biological discovery by providing highly specific and unbiased biological hypotheses for experimental testing.

Conflicts of interest

The authors declare no conflicts of interest.

References

1. Hieter, P. & M. Boguski. 1997. Functional genomics: it's all how you read it. *Science* **278:** 601–602.
2. Vaske, C.J. *et al.* 2010. Inference of patient-specific pathway activities from multi-dimensional cancer genomics data using PARADIGM. *Bioinformatics* **26:** i237–i245.
3. Louie, B., P. Mork, F. Martin-Sanchez, *et al.* 2007. Data integration and genomic medicine. *J. Biomed. Inform.* **40:** 5–16.
4. Fraser, A.G. & E.M. Marcotte. 2004. A probabilistic view of gene function. *Nat. Genet.* **36:** 559–564.
5. Pena-Castillo, L. *et al.* 2008. A critical assessment of Mus musculus gene function prediction using integrated genomic evidence. *Genome Biol.* **9**(Suppl 1): S2.
6. Mostafavi, S., D. Ray, D. Warde-Farley, *et al.* 2008. GeneMANIA: a real-time multiple association network integration algorithm for predicting gene function. *Genome. Biol.* **9** Suppl 1: S4.
7. Hart, G.T., I. Lee & E.R. Marcotte. 2007. A high-accuracy consensus map of yeast protein complexes reveals modular nature of gene essentiality. *BMC Bioinformatics* **8:** 236.
8. Warde-Farley, D. *et al.* 2010. The GeneMANIA prediction server: biological network integration for gene prioritization and predicting gene function. *Nucleic Acids Res.* **38:** W214–W220.
9. Setlur, S.R. *et al.* 2007. Integrative microarray analysis of pathways dysregulated in metastatic prostate cancer. *Cancer Res.* **67:** 10296–10303.
10. Lee, I. *et al.* 2008. A single gene network accurately predicts phenotypic effects of gene perturbation in Caenorhabditis elegans. *Nat. Genet.* **40:** 181–188.
11. Myers, C.L., D.R. Barrett, M.A. Hibbs, *et al.* 2006. Finding function: evaluation methods for functional genomic data. *BMC Genomics* **7:** 187.
12. Ashburner, M. *et al.* 2000. Gene ontology: tool for the unification of biology. The gene ontology consortium. *Nat. Genet.* **25:** 25–29.
13. Reference Genome Group of the Gene Ontology Consortium. 2009. The Gene Ontology's Reference Genome Project: a unified framework for functional annotation across species. *PLoS Comput. Biol.* **5:** e1000431.
14. Kanehisa, M. *et al.* 1999. KEGG: Kyoto Encyclopedia of Genes and Genomes. *Nucleic Acids Res.* **27:** 29–34.
15. Huttenhower, C. *et al.* 2009. The impact of incomplete knowledge on evaluation: an experimental benchmark for protein function prediction. *Bioinformatics* **25:** 2404–2410.
16. Hibbs, M.A. *et al.* 2009. Directing experimental biology: a case study in mitochondrial biogenesis. *PLoS Comput. Biol.* **5:** e1000322.
17. Gillis, J. & P. Pavlidis. 2011. The impact of multifunctional genes on "guilt by association" analysis. *PLoS One* **6:** e17258.
18. King, R.D., A. Karwath, A. Clare, *et al.* 2000. Accurate prediction of protein functional class from sequence in the *Mycobacterium tuberculosis* and *Escherichia coli* genomes using data mining. *Yeast* **17:** 283–293.
19. Kasif, S. *et al.* 2004. Whole-genome annotation by using evidence integration in functional-linkage networks. *P. Natl. Acad. Sci. USA* **101:** 2888–2893.
20. Massjouni, N., C.G. Rivera & T.M. Murali. 2006. VIRGO: computational prediction of gene functions. *Nucleic Acids Res.* **34:** W340–W344.
21. Chen, Y., Z. Li & J. Liu. 2009. *Learning Kernel Matrix from Gene Ontology and Annotation Data for Protein Function Prediction Advances in Neural Networks–ISNN 2009.* Vol. **5553** W. Yu, H. He & N. Zhang, Eds.: 694–703. Springer. Berlin/Heidelberg.

22. Lippert, C., Z. Ghahramani & K.M. Borgwardt. 2010. Gene function prediction from synthetic lethality networks via ranking on demand. *Bioinformatics* **26:** 912–918.

23. Kourmpetis, Y.A., A.D. van Dijk, M.C. Bink, *et al.* 2010. Bayesian Markov Random Field analysis for protein function prediction based on network data. *PLoS One* **5:** e9293.

24. Lord, P.W., R.D. Stevens, A. Brass, *et al.* 2003. Investigating semantic similarity measures across the gene ontology: the relationship between sequence and annotation. *Bioinformatics* **19:** 1275–1283.

25. Hess, D.C. *et al.* 2009. Computationally driven, quantitative experiments discover genes required for mitochondrial biogenesis. *PLoS Genet* **5:** e1000407.

26. Kasif, S. *et al.* 2011. COMBREX: a project to accelerate the functional annotation of prokaryotic genomes. *Nucleic Acids Res.* **39:** D11–D14.

27. Gerstein, M., M. Seringhaus & S. Fields. 2007. Structured digital abstract makes text mining easy. *Nature* **447:** 142.

28. Ripple, A.M., J.G. Mork, L.S. Knecht, *et al.* 2011. A retrospective cohort study of structured abstracts in MEDLINE, 1992–2006. *J. Med. Libr. Assoc.* **99:** 160–163.

29. Greene, C.S. & O.G. Troyanskaya. 2011. PILGRM: an interactive data-driven discovery platform for expert biologists. *Nucleic Acids Res.* **39:** W368–W374.

Ann. N.Y. Acad. Sci. ISSN 0077-8923

ANNALS OF THE NEW YORK ACADEMY OF SCIENCES
Issue: *Blavatnik Awards for Young Scientists*

Molecular dispersion spectroscopy – new capabilities in laser chemical sensing

Michal Nikodem and Gerard Wysocki

Electrical Engineering Department, Princeton University, Princeton, New Jersey

Address for correspondence: Gerard Wysocki, Electrical Engineering Department, Princeton University, Olden St. B324 EQuad, Princeton, NJ 08544. gwysocki@princeton.edu

Laser spectroscopic techniques suitable for molecular dispersion sensing enable new applications and strategies in chemical detection. This paper discusses the current state of the art and provides an overview of recently developed chirped laser dispersion spectroscopy (CLaDS)–based techniques. CLaDS and its derivatives allow for quantitative spectroscopy of trace gases and enable new capabilities, such as extended dynamic range of concentration measurements, high immunity to photodetected intensity fluctuations, or capability of direct processing of spectroscopic signals in optical domain. Several experimental configurations based on quantum cascade lasers and examples of molecular spectroscopic data are presented to demonstrate capabilities of molecular dispersion spectroscopy in the mid-infrared spectral region.

Keywords: molecular dispersion spectroscopy; laser spectroscopy; quantum cascade lasers

Introduction

Spectroscopic analysis is one of the most powerful tools used for the study of physical, quantum, and chemical properties of matter. The accuracy and precision of spectroscopic analysis benefited tremendously from the application of lasers that were first experimentally demonstrated in 1960,[1] and since then they have been extensively used as spectroscopic sources. Laser parameters such as highly monochromatic radiation, high spectral brightness, short laser pulses, or wavelength tunability have enabled a wide variety of applications, ranging from chemical identification and quantification,[2] through studies of nonlinear optical phenomena,[3] to precise probing of ultra-fast chemical processes.[4] Among all laser-based spectroscopic techniques (such as laser-induced fluorescence, laser photoaccoustics, and Raman spectroscopy), laser absorption spectroscopy (LAS) and its derivatives have been by far the most popular and widely used. LAS became popular mainly due to its high sensitivity (down to pptv, parts-per-trillion by volume),[5] high specificity, and a relatively simple setup. Bandwidth-normalized absorption detection lim-

its between $10^{-4}/Hz^{1/2}$ and $10^{-6}/Hz^{1/2}$ for the best LAS systems have been reported.[5] LAS spectrometers have been successfully used for chemical sensing both in laboratories as well as in the field,[6,7] especially after compact electrically pumped semiconductor laser sources have become available.[8] The physics of LAS is well understood (governed by the Beer-Lambert law), which provides a convenient tool for modeling of the experimental results obtained in various configurations, including *in situ* nondestructive sensing[9–11] or remote detection schemes.[12]

The implementation of LAS is relatively easy and requires a measurement of the laser power absorbed within a sample. Since this is usually performed by measurement of the total laser intensity transmitted through the sample, some limitations of this process become significant, especially in the systems that target high accuracy absorption measurements. The major limitations include nonlinearity of the Beer-Lambert law, especially for strongly absorbing samples (beyond 10–30% absorption), dynamic range limitations due to quantification of small changes in the total photo-detected laser intensity or due to nonlinearities of the photodetectors, and direct

doi: 10.1111/j.1749-6632.2012.06660.x

impact of the intensity noise on the measured absorption signal. All those limitations are common to any detection scheme that relies on direct intensity measurements.

Since the absorption process also affects the phase of the electromagnetic radiation transmitted through the sample (dispersion), instead of LAS measurements, one can perform phase measurements to retrieve the same spectral information about the sample. Fundamentally, knowledge of the sample's absorption coefficient at all frequencies allows the determination of the sample dispersion by applying the Kramers–Kronig relations.[13] By using an approximation for a weakly absorbing sample the Kramers–Kronig transformation of the absorption coefficient $\alpha(\omega)$ into the sample's refractive index $n(\omega)$ can be expressed as:

$$n(\omega) = 1 + \frac{c}{\pi} \int_0^{+\infty} \frac{\alpha(\omega')}{\omega'^2 - \omega^2} d\omega', \qquad (1)$$

where c is the speed of light and ω the optical angular frequency. This allows a measurement of the refractive index to be modeled using absorption spectra databases.

Experimental approaches that considered detection of the refractive index of the sample in the vicinity of a molecular or atomic transition were studied a century ago using only prisms and grating spectrometers.[14] Using coherent radiation, the sensitivity of those measurements can be significantly improved, usually at the cost of higher complexity of the detection systems.[15] Although more challenging than conventional absorption measurements, the measurement of the refractive index (phase) can offer many potential advantages over absorption sensing. The most significant include the linear relationship between the dispersion spectrum and the sample concentration (unlike absorption that saturates with increasing concentration), phase measurements that are immune to amplitude/intensity fluctuations, and effects of the photodetector nonlinearity that are effectively mitigated in measurements that rely on phase detection instead of amplitude.

Example methods capable of dispersion sensing

Most of the currently available measurement methods that give access to refractive index information are based on coherent laser sources and interferometric detection. There are also some techniques that exploit measurement of natural or artificially created sample birefringence, which effectively would classify them as dispersion-based methods. In the following sections a brief literature overview of four different measurement schemes that give access to, or rely on, sample dispersion measurement will be provided.

Frequency modulation spectroscopy

One of the derivative methods of LAS employs high-frequency modulation of the laser source and gives access to both absorption and dispersion information. The method known as frequency modulation (FM)-spectroscopy was first proposed by Bjorklund[16] and requires application of high-frequency modulation at ω_m applied to the laser radiation to produce FM sidebands around the optical carrier. If there is no absorption line coinciding with the carrier frequency, the beatnotes originating from the heterodyning between each sideband and the carrier cancel out and give no heterodyne signal at ω_m. However, if the wavelength of the laser is tuned across the target transition, the sideband symmetry is perturbed, which results in a measurable heterodyne signal (see Fig. 1A). Both absorption and dispersion are represented by the sample's complex transmission function and can be de-convolved from the demodulated heterodyne beatnote by selecting an appropriate detection phase. FM-spectroscopy shifts the detection to the high-frequency region, which helps avoid 1/f noise. It is a very sensitive method for absorption measurement, with bandwidth-normalized absorption detection limits reported below $10^{-6}/\text{Hz}^{1/2}$ (Ref. 17), but the dispersion information is rarely used for molecular sensing. In a typical implementation of this technique both the sample's absorption and dispersion are detected through intensity measurement (as shown in Fig. 1A); thus both measurements are affected by some fundamental limitations that are characteristic for conventional LAS (e.g., performance depends on intensity fluctuations and amplitude noise). Moreover, this technique can only be applied when sample absorption/dispersion is small (see assumptions in Ref. 18); therefore, it does not take full advantage of capabilities that are offered by the dispersion sensing (i.e., linear response for a wide range of concentrations).

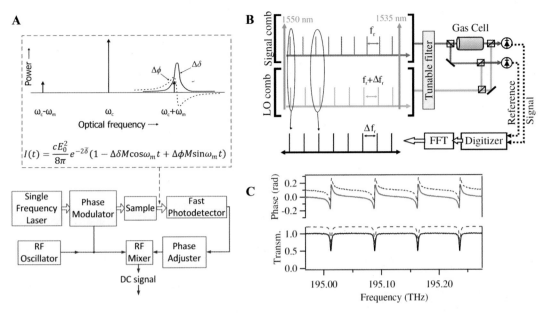

Figure 1. (A) An experimental arrangement for FM spectroscopy with its frequency domain illustration (from Ref. 18). $\Delta\delta$ and $\Delta\phi$ represent attenuation (due to absorption) and phase shift (due to dispersion), respectively, of the sideband at $\omega_c+\omega_m$ with respect to carrier at ω_c; $\bar{\delta}$ denotes attenuation of the sideband at $\omega_c-\omega_m$ with respect to carrier at ω_c and is assumed to be constant. Absorption and dispersion spectra are retrieved from the light intensity $I(t)$ as its in-phase and quadrature components (see details in Refs. 16 and 18); (B) an experimental layout for dual-comb spectroscopy (from Ref. 27). A signal comb and a local oscillator (LO) comb are phase locked and have slightly different repetition rates. Beating them with each other enables down-converting of the optical information carried by each signal comb tooth to the distinct RF frequency. Absorption and optical phase change can be retrieved from the amplitude and the phase of the RF beatnotes; (C) absorption and refractive index spectra of acetylene measured using dual-comb spectrometer (from Ref. 27).

NICE-OHMS

A special case of FM-spectroscopy that also gives access to dispersion information is noise-immune cavity-enhanced optical heterodyne molecular spectrometry (NICE-OHMS). This technique is essentially a combination of FM-spectroscopy with a high Finesse cavity-enhanced absorption spectroscopy. Some recent reports on implementation of NICE-OHMS to molecular spectroscopy in the near infrared (near-IR) and the mid-IR spectral regions have been published in Refs. 19 and 20, respectively. NICE-OHMS requires a rather complex optical system with extremely precise stabilization of multiple active and passive optical components, which makes it impossible to implement outside the laboratory. However, because of a long effective optical path-length within the sample and high immunity to laser noise, NICE-OHMS spectrometers achieve ultra-high sensitivities (with minimum absorption coefficient detection limits down to 10^{-14} per cm). Additionally, since NICE-OHMS essentially uses an FM-spectroscopy approach for signal demodula-

tion, the extraction of the dispersion signal is prone to the same limitations as the FM-spectroscopy.

Spectroscopy with optical frequency combs

Another group of recently developed methods that allow extraction of information about the sample dispersion are techniques employing optical frequency combs (OFCs). OFCs are pulsed laser sources that emit broadband radiation composed of narrow phase-locked spectral emission lines (comb teeth) that are equidistantly spaced in the frequency domain with spacing dictated by the reciprocal of the pulse repetition interval. Because of the mode-locked character of the generated light, this pulsed radiation can be considered as a superposition of continuous-wave (cw) frequency teeth components that can be used for coherent detection and extraction of optical phase information. A variety of measurement schemes have been proposed to date, including spectrometers based on dispersion gratings and other dispersive elements,[21] dual-comb heterodyne detection,[22] or OFC-based

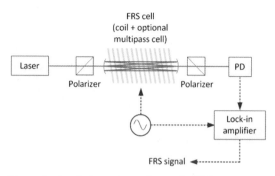

Figure 2. A typical experimental set-up for FRS.

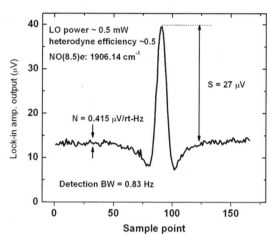

Figure 3. Spectrum of NO R(8.5)e transition acquired with heterodyne-enhanced FRS. Experimental conditions: NO concentration, 2 ppm-v; sample pressure, 30 Torr; sample temperature, 300 K; magnetic field, ∼100 Gauss; optical path length in Faraday cell, L = 15 cm; optical power before analyzer, $P_0 = 14$ mW; local oscillator (LO) power, $P_{LO} = 0.5$ mW; and detection bandwidth of 0.83 Hz.[30]

Fourier transform spectroscopy.[23] Multiple coherent detection schemes for applications in molecular spectroscopy have been reported (for example, implementation using dual-comb heterodyne detection together with a general concept of operation principle are shown in Figs. 1B and C). OFC-based techniques provide very high-frequency precision and bandwidth-normalized absorption detection limits at the $10^{-4}/Hz^{1/2}$ level (e.g., in the cavity-enhanced OFC spectroscopy reported in Ref. 24, the noise estimated in the cavity transmission spectrum was at the level of $7.1 \times 10^{-4}/Hz^{1/2}$, which translates to a minimum detectable absorption coefficient of 5.4×10^{-9} per cm per $Hz^{1/2}$ when cavity enhancement is taken into account). Coherent detection schemes also give access to the sample's dispersion information,[25–27] but similarly to FM spectroscopy, the dispersion detection is rarely used for sensing. Moreover, owing to extremely high complexity of the optical setups, their applications are limited only to highly specialized optical laboratories.

Faraday rotation spectroscopy

One spectroscopic method that relies on molecular dispersion measurement and that can be used outside specialized laboratories is Faraday rotation spectroscopy (FRS). FRS is one of the special cases of dispersion sensing that uses magnetically created circular birefringence of the sample to induce the Faraday effect.[28,29] The Faraday effect manifests itself as the rotation of polarization axis of linearly polarized light transmitted through the sample, which is subsequently measured in FRS systems. FRS requires magneto-optical activity of the sample and thus can only be applied to paramagnetic species. The polarization rotation measurement is essentially a measurement of optical phase shift

that originates from a difference between refractive indices for the right-handed and left-handed circularly polarized wave components propagating through the sample. Thus effectively, the FRS is a dispersion-based measurement that provides many valuable features that are not available with standard absorption-based techniques.

We have studied FRS extensively[28–31] and demonstrated several transportable and portable autonomous cryogen-free FRS systems, both in near-IR and mid-IR. A typical schematic of an optical setup for FRS is shown in Figure 2. In the FRS system, employing a direct photodetection with a single detector element, the amount of power transmitted by the second (nearly crossed) polarizer is usually very small. Therefore, further enhancement of the signal-to-noise ratio (SNR) can be achieved by applying more sensitive photodetection techniques. After application of a heterodyne-enhanced detection scheme the photodetection sensitivity in the FRS system was significantly improved to the levels that are close to the fundamental shot noise limit.[30] A spectrum of nitric oxide (NO) acquired with the heterodyne-enhanced FRS system operating at 5.25 μm is shown in Figure 3. Analysis of the SNR provides the noise equivalent detectable polarization rotation angle of $2.6 \times 10^{-8} \frac{rad}{\sqrt{Hz}}$, which, for this particular system (15 cm total optical path),

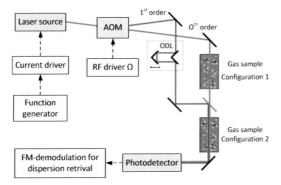

Figure 4. Optical layout for CLaDS (AOM, acousto-optical modulator; RF, radio frequency; ODL, optical delay line).

corresponds to the difference in refractive index for right- and left-handed polarizations at the level of $\sim 1.5 \times 10^{-13}\,\mathrm{Hz}^{-1/2}$. Besides ultra-high sensitivity (e.g., an effective bandwidth-normalized absorption detection limit of $5.9 \times 10^{-8}\,\mathrm{Hz}^{-1/2}$ was reported in Ref. 29), FRS also provides a high dynamic range of concentration measurements as expected for dispersion-based sensing technology. It was shown in Ref. 28 that concentration measurements, ranging from a single ppbv up to 10 ppmv levels of NO, could be detected with no signal saturation and with no impact on the actual precision of the FRS system.

Direct measurement of molecular refractive index and applications

As shown earlier, dispersion sensing has a potential for sensitive and high dynamic range chemical detection. However, the existing methods that are capable of dispersion sensing show some limitations, and none of them is truly comparable to conventional LAS for simplicity, functionality, and ease of application. For example, the cost and complexity of frequency comb spectroscopy is too high to use for basic trace-gas sensing; applications of NICE-OHMS are practically limited to laboratory experiments; FM spectroscopy relies on light intensity detection that carries limitations of standard absorption-based techniques; and FRS, although very practical, can only be applied to paramagnetic species. Therefore a new sensing scheme that is capable of measuring molecular dispersion with relatively simple and robust setup has a potential to overcome some of these drawbacks and provide advantages of dispersion spectroscopy. Such

attempts have been made in the past, but only a small amount of progress has been made in adoption of proposed techniques to sensitive trace-gas detection in nonlaboratory conditions.[15,19,32–37] Recently we have introduced a new chirped laser dispersion spectroscopy (CLaDS) technique that relies on molecular dispersion sensing in the gas sample,[38] and the system itself can be easily adopted for field measurements.[39] The following sections provide details about fundamentals, unique capabilities, and potential applications of CLaDS and related technologies.

Direct CLaDS
The schematic diagram of the CLaDS system is shown in Figure 4. It uses a single-frequency laser source that is frequency chirped across the molecular transition of interest. Laser radiation is directed through an acousto-optical modulator (AOM) that shifts the frequency of light by Ω. Two beams (the fundamental 0th order beam and the frequency-shifted 1st order beam) are combined into one dual-color beam using Mach-Zehnder arrangement. The dual-color beam is focused on a fast photodetector, and a heterodyne beatnote between the two waves is measured and analyzed. The measured sample can be located either inside the interferometer ("Configuration 1") or after the beam combiner ("Configuration 2"). Configuration 2 is also suitable for multi-pass cell arrangements or remote long-distance open-path sensing. The dispersion in the sample gas has a slightly different effect on the propagation of the two light waves. It results in a difference in propagation times Δt for the two frequency-shifted waves, which affects the frequency of the measured heterodyne beatnote and changes it by $S \cdot \Delta t$, where S is the chirp rate that additionally enhances the dispersion signal. By performing frequency-demodulation of the beatnote while the laser is chirped, a dispersion spectrum encoded into the frequency of the heterodyne beatnote can be retrieved and given by:[38]

$$f(\omega) = \frac{1}{2\pi}\left[\Omega + \frac{S \cdot \Delta L}{c} - \frac{S \cdot L_c}{c} \cdot \omega \right.$$
$$\left. \cdot \left(\frac{dn}{d\omega}\bigg|_{\omega - \Omega} - \frac{dn}{d\omega}\bigg|_{\omega}\right)\right], \qquad (2)$$

where ΔL is the path length difference between the Mach-Zehnder interferometer arms, L_c is the path-length within the sample, and n is the refractive

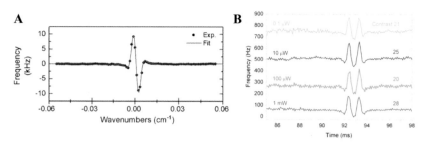

Figure 5. (A) An example of a model CLaDS spectrum fitted to the data recorded with a direct CLaDS detection scheme. Line-by-line spectral calculation based on the HITRAN database was used to model the NO spectrum around 1,906 cm^{-1}. (B) Molecular dispersion spectra measured in Configuration 1 for a wide range of detected RF powers (spectra are shifted vertically for viewing purposes; RF power is shown next to each spectrum). Signal amplitude is unaffected, even when RF beatnote power changes by four orders of magnitude (measured transitions is NO doublet at 1,912.08 cm^{-1}).[38]

index of the medium. The amplitude of the measured dispersion line (third term in Eq. 2) is proportional to the chirp rate S and to the spectral derivative of the sample refractive index. The latter is proportional to the sample concentration, which allows for quantitative chemical detection.

The beatnote is frequency demodulated at the carrier Ω (which is typically in radio frequency range) and in addition to optical dispersion information about ΔL can also be retrieved. In the case of molecular dispersion sensing the contribution due to path difference, ΔL is excessive and can be suppressed by balancing the arms of the Mach-Zehnder interferometer (e.g., by using an optical delay line). Hence, when $\Delta L = 0$, the FM-demodulation provides virtually a baseline-free information on molecular dispersion. This is the simplest and the most straightforward way of direct signal recording in CLaDS (direct-CLaDS).

Because of nonlinear frequency chirping (which is typically observed with both cw and pulsed laser sources), direct-CLaDS requires precise chirp characterization before quantitative data can be recorded. This can be performed either with an etalon inserted into the optical path and subsequent fringe analysis or by setting $\Delta L \neq 0$, which enables recording of a baseline signal that is proportional to the chirp rate.[38] When the chirp rate is known, the recorded dispersion spectrum can be fitted with a model spectrum in order to retrieve parameters of the sample. Since the relationship between the sample absorbance and its dispersion is known (see Eq. 1), direct CLaDS provides the same modeling capabilities as conventional absorption spectroscopy.

Spectral information related to concentration, pressure broadening, or temperature are all encoded in the profile of the dispersion feature and can be retrieved using spectroscopic modeling. An example of such a dispersion fitting that uses parameters from the HITRAN database and the plasma dispersion function (Voigt line profile) for calculation of the CLaDS spectrum is shown in Figure 5. For this experiment a Quantum Cascade Laser (QCL), operating at 5.2 μm was used to target ro-vibrational transition of NO. A gas sample (850 ppm of NO in N_2, total pressure of 6 Torr) contained within a 10 cm–long cell was placed in the Configuration 2 arrangement. By spectral fitting of the recorded data, the concentration of NO and the total sample gas pressure were retrieved. Through precise balancing of the interferometer arms ($\Delta L = 0$), a baseline-free CLaDS spectrum was recorded. This is a significant advantage over the direct LAS detection that requires extracting useful spectroscopic signal from a high and often fluctuating intensity background. Other important properties of the signal in CLaDS are a linear dependence of its amplitude on concentration[40] and its immunity to changes in photodetected intensity. Since the dispersion spectrum is retrieved through FM-demodulation (measurement of the heterodyne beatnote frequency), the intensity (amplitude) has no impact on the measured spectroscopic signal and has only a minor effect on the SNR. As shown in Figure 5, the peak-to-peak amplitude of the CLaDS spectrum is not affected, and the SNR remains nearly the same even in the presence of significant (few orders of magnitude) changes in the RF beatnote power (see Ref. 38 for details).

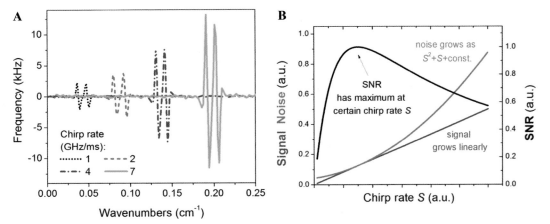

Figure 6. (A) Four spectra showing NO doublet around 1,905.2 cm^{-1}, recorded for different chirp rates (sample pressure was 10 Torr; spectra are offset laterally for viewing purposes). As expected from Eq. 1, the CLaDS signal amplitude is proportional to the chirp rate. (B) General dependence of the signal and noise on the chirp rate results in an optimum chirp rate that is required to achieve the best SNR.

An important challenge in direct-CLaDS is selection of an optimum chirp rate that will provide the highest possible SNR. The measured CLaDS signal can be additionally enhanced by increasing the chirp rate S, which, in the case of semiconductor lasers, can easily reach values above 10^{15} Hz/sec. Unfortunately, as shown in Figure 6, an increase of S, in addition to signal "amplification" ($\propto S$), also results in an increase of noise ($\propto S^2 + S + \text{const.}$). As a consequence there is an optimum chirp rate for which maximum SNR occurs in direct-CLaDS measurement. This optimum chirp rate depends on the relative contribution from different sources of noise, mainly from optical fringes and from FM-demodulation noise. Therefore an individual optimization should be performed for each particular CLaDS system (detailed analysis of SNR in CLaDS can be found in Ref. 41).

Chirped-modulated CLaDS

Direct-CLaDS, although relatively simple in experimental implementation and modeling, shows some important limitations. In order to provide a stable baseline-free signal, it requires additional mechanical stabilization of the optical setup. Otherwise, vibrations and mechanical drifts that create an imbalance of the interferometer ($\Delta L \neq 0$) can produce uncontrolled baseline fluctuations in the measured spectrum. Moreover, in the direct-CLaDS laser, frequency needs to be chirped across the frequency range that is significantly wider than the measured

transition width to capture potential baseline drifts. This might be challenging in the case of open-path sensing when typical linewidths are larger than 3 GHz. In order to increase sensitivity of CLaDS and enable true baseline-free detection, even in atmospheric conditions (i.e., open path measurement), we have developed a new detection scheme based on chirp modulation. In a chirp-modulated CLaDS (CM-CLaDS), laser current is modulated with a sinusoidal signal at frequency f, which produces sinusoidal modulation of the chirp rate. The dispersion spectrum is also encoded in the frequency of the beatnote, but only harmonics of the chirp modulation frequency (i.e., $N \times f$, $N = 1, 2, 3\ldots$) need to be analyzed. Because of that, the FM-demodulation bandwidth can be significantly reduced, which results in lower FM-demodulation noise. Moreover, even if the imbalance of the interferometer exists ($\Delta L \neq 0$), it only produces baseline in the $1f$ spectrum. Therefore, by targeting higher harmonics, the baseline is suppressed. The most desirable harmonics are $3f$ for Configuration 1 or $2f$ for Configuration 2. In both cases, the CM-CLaDS signal reaches a maximum at the line center, and its amplitude can be directly converted into targeting an analyte concentration after initial calibration of the system. In Figure 7A, a sample $2f$ spectrum acquired in Configuration 2 is presented. It was recorded in an open-path arrangement with 4.54 μm QCL used as a spectroscopic source to perform detection of nitrous oxide (N_2O). A laser beam from the CLaDS

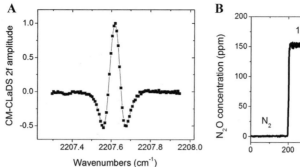
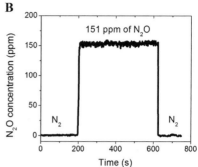

Figure 7. (A) A 2f CM-CLaDS spectrum of N_2O line measured in a remote sensing arrangement (Configuration 2). Experimental conditions: ambient air, total optical path of 70 m. (B) Continuous monitoring of N_2O concentration was performed by recording a CM-CLaDS 2f signal at the N_2O line center (experimental conditions: 100 Torr pressure, 10 cm optical path).

setup was directed toward a distant (placed 35 m away) retroreflector, and the returning light was collected on a photodetector. The 2f CM-CLaDS spectrum of N_2O is indeed baseline free, which allows for continuous concentration monitoring, provided the laser is locked to the center of the target transition.

Shown in Figure 7B is an example of a time series recorded for continuous monitoring of N_2O concentration. The peak of the 2f CM-CLaDS signal amplitude at line center was measured continuously, and two gas samples (N_2O/N_2 mixture or dry nitrogen) were alternately flown through a 10 cm–long optical gas cell. Preliminary results show that by reducing FM-demodulation noise, the CM-CLaDS technique can provide at least one order of magnitude improvement of the instrument sensitivity with respect to direct-CLaDS. Moreover, it enables concentration monitoring with no need for real-time fitting of the recorded spectra. For the prototype sensor (with a 15-cm sensing path), we have determined the minimum detection limit to N_2O of 120 ppbv-m/Hz$^{1/2}$, which corresponds to an equivalent minimum detectable fractional absorption of 2.4×10^{-4} Hz$^{-1/2}$. This is within a range of performance reported for conventional LAS systems and OFC-based spectrometers. Since CLaDS is in the very early stage of its development, further improvements, such as suppression of parasitic fringe noise,[41] or generation of optimal frequency shift,[38] are expected to provide sensitivity enhancement, improved stability of the system,[42] and reduction of system complexity. Up to now, the relatively small complexity of the CLaDS system, ease of application, ability to measure optically thick samples, and immunity to intensity/transmission fluctuations have already proved to be helpful in overcoming the biggest drawbacks of the standard absorption-based methods or FM-spectroscopy, and allowed first field deployments of the prototype system.[43]

Differential optical dispersion spectroscopy

Dispersion spectroscopy, due to its unique properties, can also provide entirely new capabilities in trace-gas detection. Differential optical dispersion spectroscopy (DODiS) is a new generation of spectroscopic measurement that builds upon the CLaDS approach and enables true optical subtraction or addition of spectroscopic signals.[44] In DODiS (optical layout shown in Fig. 8A), two gas samples are separately placed in the two arms of the interferometer. In one arm the laser beam first passes through one gas sample and then is frequency shifted and combined with the second beam (that interacts with the second gas sample) on a beamsplitter. The combined beams are focused on the photodetector. Similarly as in CLaDS, an FM-demodulation of the heterodyne beatnote is performed in order to retrieve the target dispersion profile, and an AM-demodulation can be used for absorption signal retrieval. Because the frequency shifter is located after the gas sample cell, both samples are probed with the light at the same optical frequency. Therefore the demodulated signal corresponds to an actual difference in propagation time through both gas cells. As a result, DODiS enables direct detection of a difference between dispersion spectra of both samples. Such a subtraction of two spectroscopic signals is performed entirely in the optical domain and does not

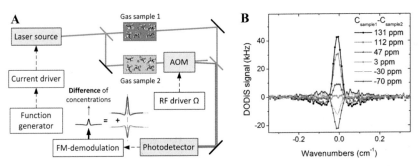

Figure 8. (A) An optical arrangement of DODiS. The AOM is placed after gas sample #2, which is a key difference with respect to CLaDS (when AOM is placed before the gas sample, the light interacting with each sample has a slightly different frequency shifted by Ω; thus dispersion spectra from the two samples measured at the beatnote frequency cannot be fully subtracted). (B) Example of differential measurement of two samples, both containing an N_2O mixture at a total pressure of 60 Torr (target transition was N_2O line at 2,209.52 cm^{-1}).[44]

require samples to be measured separately. At the same time, the AM-demodulated absorption spectra are additive, which provides optical summing.

In Figure 8B, an example of N_2O measurement using DODiS is presented. In this experiment the gas cell #2 was filled with approximately 90 ppm of N_2O (balanced with nitrogen). The concentration in cell #1 was varied (from more than 200 ppm to almost zero), and the DODiS dispersion signal was measured and processed. The recorded spectra were fitted using the HITRAN database, and information on concentration difference ($C_{sample1} - C_{sample2}$) was retrieved. DODiS enables direct detection of concentration difference, which can be particularly useful in all kinds of measurements where two gas samples need to be compared, for example, in isotopic ratio measurement. Because the amount of molecular dispersion changes linearly with molecular concentration, differential detection with DODiS can be performed even for highly absorbing (>10–30% absorption) samples. This enables increased sensitivities using extended optical paths, which might cause saturation effects if standard absorption-based methods are used. DODiS also allows for selective cancellation/suppression of unwanted background that can interfere with the spectral feature of the target analyte.[44]

Conclusions

For several decades laser-based trace-gas sensing instrumentation has been shown to provide highly sensitive and selective chemical detection. However, in order to meet challenging demands of new ap-

plications that require even higher dynamic ranges, improved stability, and increased robustness, new detection schemes are needed. As discussed in this paper, recent developments in the area of molecular dispersion sensing are very promising and show potential for overcoming the drawbacks of conventional technologies while preserving most of their advantages, which include simplicity and ease of operation. Specifically, CLaDS translates the spectroscopic detection from intensity/amplitude domain to a measurement performed in the frequency domain, which provides increased stability and immunity to amplitude noise. It offers multiple new sensing capabilities that are not accessible through traditional techniques, such as linear response and a higher dynamic range of concentration measurements. In addition to improvements with respect to existing LAS systems, the molecular dispersion measurement provides unique advantages of spectroscopic signal processing directly in the optical domain (i.e., spectral addition or subtraction), which has been demonstrated with DODiS. In this paper, we have presented data that should demonstrate some of the unique capabilities of spectroscopic techniques that rely on dispersion measurement. The first prototypes of CLaDS-based instruments have already been used in early field deployments,[43] which indicate the technological feasibility for real-world molecular sensing applications.

Acknowledgments

The authors acknowledge the financial support from NSF CAREER Award CMMI-0954897 and by Grants from the National Center for Research

Resources (5R21RR026231–03) and the National Institute of General Medical Sciences (8 R21 GM103460–03), the National Institutes of Health. Dr. Chung-En Zah at Corning Inc. and Dr. Antoine Muller at Alpes Lasers SA are acknowledged for providing lasers for this study.

Conflicts of interest

The authors declare no conflicts of interest.

References

1. Maiman, T.H. 1960. Stimulated optical radiation in ruby. *Nature* **187:** 493–494.
2. McManus, J.B. *et al.* 2010. Application of quantum cascade lasers to high-precision atmospheric trace gas measurements. *Opt. Eng.* **49:** 111124–111124-11.
3. Dreier, T. *et al.* 1988. Determination of temperature and concentration of molecular nitrogen, oxygen and methane with coherent anti-stokes raman scattering. *Appl. Phys. B: Lasers Opt.* **45:** 183–190.
4. Zewail, A.H. 1988. Laser femtochemistry. *Science* **242:** 1645–1653.
5. Richter, D. *et al.* 2009. Difference frequency generation laser based spectrometers. *Laser Photon. Rev.* **3:** 343–354.
6. Webster, C.R. *et al.* 2004. Mars Laser Hygrometer. *Appl. Opt.* **43:** 4436–4445.
7. Wysocki, G. *et al.* 2007. Dual interband cascade laser based trace-gas sensor for environmental monitoring. *Appl. Opt.* **46:** 8202–8210.
8. Werle, P. *et al.* 2002. Near- and mid-infrared laser-optical sensors for gas analysis. *Opt. Lasers Eng.* **37:** 101–114.
9. Wysocki, G. *et al.* 2004. Pulsed quantum-cascade laser-based sensor for trace-gas detection of carbonyl sulfide. *Appl. Opt.* **43:** 6040–6046.
10. Wysocki, G. *et al.* 2005. Exhaled human breath analysis with quantum cascade laser based gas sensors. In *Breath Analysis for Clinical Diagnosis and Therapeutic Monitoring*. A. Amann and D. Smith, Eds. World Scientific. Singapore.
11. McCurdy, M.R. *et al.* 2007. Recent advances of laser-spectroscopy-based techniques for applications in breath analysis. *J. Breath Res.* **1:** 014001.
12. Weibring, P. *et al.* 2004. Multi-component chemical analysis of gas mixtures using a continuously tuneable lidar system. *Appl. Phys. B: Lasers Opt.* **79:** 525–530.
13. Demtroder, W. 1981. *Laser Spectroscopy: Basic Concepts and Instrumentation/Wolfgang Demtroder*. Springer-Verlag. Berlin, New York.
14. Wood, R.W. 1901. The anomalous dispersion of sodium vapour. *Proc. Royal Soc. London* **69:** 157–171.
15. Denchev, O.E. *et al.* 1982. Possibility of intra-resonator double-beam spectrointerferometry of phase objects using a dye laser. *J. Appl. Spectrosc.* **36:** 267–271.
16. Bjorklund, G.C. 1980. Frequency-modulation spectroscopy: a new method for measuring weak absorptions and dispersions. *Opt. Lett.* **5:** 15–17.
17. Carlisle, C.B. *et al.* 1989. Quantum noise-limited FM spectroscopy with a lead-salt diode laser. *Appl. Opt.* **28:** 2567–2576.
18. Bjorklund, G.C. *et al.* 1983. Frequency modulation (FM) spectroscopy. *Appl. Phys. B: Lasers Opt.* **32:** 145–152.
19. Foltynowicz, A. *et al.* 2008. Characterization of fiber-laser-based sub-Doppler NICE-OHMS for quantitative trace gas detection. *Opt. Exp.* **16:** 14689–14702.
20. Taubman, M.S. *et al.* 2004. Stabilization, injection and control of quantum cascade lasers, and their application to chemical sensing in the infrared. *Spectrochimica Acta Part A: Mol. Biomol. Spectrosc.* **60:** 3457–3468.
21. Thorpe, M.J. *et al.* 2008. Cavity-enhanced optical frequency comb spectroscopy: application to human breath analysis. *Opt. Exp.* **16:** 2387–2397.
22. Schiller, S. 2002. Spectrometry with frequency combs. *Opt. Lett.* **27:** 766–768.
23. Mandon, J. *et al.* 2009. Fourier transform spectroscopy with a laser frequency comb. *Nat. Photon* **3:** 99–102.
24. Foltynowicz, A. *et al.* 2012. Cavity-enhanced optical frequency comb spectroscopy in the mid-infrared application to trace detection of hydrogen peroxide. *Appl. Phys. B: Lasers Opt.*, DOI: 10.1007/s00340-012-5024-7
25. Newbury, N.R. *et al.* 2010. Sensitivity of coherent dual-comb spectroscopy. *Opt. Exp.* **18:** 7929–7945.
26. Mandon, J. *et al.* 2007. Frequency-modulation Fourier transform spectroscopy: a broadband method for measuring weak absorptions and dispersions. *Opt. Lett.* **32:** 2206–2208.
27. Coddington, I. *et al.* 2008. Coherent multiheterodyne spectroscopy using stabilized optical frequency combs. *Phys. Rev. Lett.* **100:** 013902.
28. Wysocki, G. *et al.* 2009. Continuous monitoring of nitric oxide at 5.33 um with an EC-QCL based Faraday rotation spectrometer: laboratory and field system performance. *Proc. SPIE* **7222:** 72220M.
29. Lewicki, R. *et al.* 2009. Ultrasensitive detection of nitric oxide at 5.33 um by using external cavity quantum cascade laser-based Faraday rotation spectroscopy. *Proc. Nat. Acad. Sci.* **106:** 12587–12592.
30. Wang, Y. *et al.* 2012. Heterodyne-enhanced Faraday rotation spectrometer. *Proc. SPIE* **8268:** 82682F.
31. So, S. *et al.* 2011. VCSEL based Faraday rotation spectroscopy with a modulated and static magnetic field for trace molecular oxygen detection. *Appl. Phys. B: Lasers Opt.* **102:** 279–291.
32. Duval, A.B. & A.I. McIntosh. 1980. Measurement of oscillator strength by tunable laser interferometry. *J. Phys. D: Appl. Phys.* **13:** 1617.
33. Gross, R. *et al.* 1980. Measurements of the anomalous dispersion of HF in absorption. *Quant. Electron., IEEE J.* **16:** 795–798.
34. Marchetti, S. & R. Simili. 2005. Measurement of the refractive index dispersion around an absorbing line. *Opt. Commun.* **249:** 37–41.
35. Moschella, J.J. *et al.* 2006. Resonant, heterodyne laser interferometer for state density measurements in atoms and ions. *Rev. Sci. Instrum.* **77:** 093108–5.
36. Werle, P. 1996. Spectroscopic trace gas analysis using semiconductor diode lasers. *Spectrochimica Acta Part A: Mol. Biomol. Spectrosc.* **52:** 805–822.
37. Schmidt, F.M. *et al.* 2010. Highly sensitive dispersion spectroscopy by probing the free spectral range of an optical cavity using dual-frequency modulation. *Appl. Phys. B: Lasers Opt.* **101:** 497–509.

38. Wysocki, G. and D. Weidmann. 2010. Molecular dispersion spectroscopy for chemical sensing using chirped mid-infrared quantum cascade laser. *Opt. Exp.* **18:** 26123–26140.

39. Wysocki, G. & M. Nikodem. 2011. Chirped laser dispersion spectroscopy for remote sensing of trace-gases. In *Renewable Energy and the Environment,* OSA Technical Digest (CD) (Optical Society of America, 2011), paper EWC1.

40. Franz, K. *et al.* 2010. High dynamic range laser dispersion spectroscopy of saturated absorption lines. In *Conference on Lasers and Electro-Optics,* Optical Society of America, p. CMJ4.

41. Nikodem, M. *et al.* 2012. Signal-to-noise ratio in chirped laser dispersion spectroscopy. *Opt. Exp.* **20:** 644–653.

42. Nikodem, M. *et al.* 2012. Chirped laser dispersion spectroscopy with harmonic detection of molecular spectra. *Appl. Phys. B: Lasers Opt.,* pp. 1–7. DOI: 10.1007/s00340-012-5060-3.

43. Nikodem, M. & G. Wysocki. 2012. Remote open-path sensing of nitrous oxide using chirped laser dispersion spectroscopy. In *Conference on Lasers and Electro-Optics,* Optical Society of America, p. CM2F.1.

44. Nikodem, M. & G. Wysocki. 2012. Differential optical dispersion spectroscopy. In *Lasers, Sources, and Related Photonic Devices,* Optical Society of America, p. LW5B.5.